Phase Modeling Tools

Chemical Thermodynamics Set

coordinated by
Michel Soustelle

Volume 1

Phase Modeling Tools

Applications to Gases

Michel Soustelle

iSTE

WILEY

First published 2015 in Great Britain and the United States by ISTE Ltd and John Wiley & Sons, Inc.

ISTE Ltd
27-37 St George's Road
London SW19 4EU
UK

www.iste.co.uk

John Wiley & Sons, Inc.
111 River Street
Hoboken, NJ 07030
USA

www.wiley.com

Library of Congress Control Number: 2015940031

British Library Cataloguing-in-Publication Data
A CIP record for this book is available from the British Library
ISBN 978-1-84821-864-2

Contents

Preface

This book – an in-depth examination of chemical thermodynamics – is written for an audience of engineering undergraduates and Masters students in the disciplines of chemistry, physical chemistry, process engineering, materials, etc., and doctoral candidates in those disciplines. It will also be useful for researchers at fundamental- or applied-research labs, dealing with issues in thermodynamics during the course of their work.

These audiences will, during their undergraduate degree, have received a grounding in general thermodynamics and chemical thermodynamics, which all science students are normally taught, and will therefore be familiar with the fundamentals, such as the principles and the basic functions of thermodynamics, and the handling of phase- and chemical equilibrium states, essentially in an ideal medium, usually for fluid phases, in the absence of electrical fields and independently of any surface effects.

This set of books, which is positioned somewhere between an introduction to the subject and a research paper, offers a detailed examination of chemical thermodynamics that is necessary in the various disciplines relating to chemical- or material sciences. It lays the groundwork necessary for students to go and read specialized publications in their different areas. It constitutes a series of reference books that touch on all of the concepts and methods. It discusses both scales of modeling: microscopic (by statistical thermodynamics) and macroscopic, and illustrates the link between them at every step. These models are then used in the study of solid, liquid and gaseous phases, either of pure substances or comprising several components.

The various volumes of the set will deal with the following topics:

– phase modeling tools: application to gases;

– phase modeling of liquid phases;

– modeling of solid phases;

– chemical equilibrium states;

– phase transformations;

– electrolytes and electrochemical thermodynamics;

– thermodynamics of surfaces, capillary systems and phases of small dimensions.

Appendices in each volume give an introduction to the general methods used in the text, and offer additional mathematical tools and some data.

This series owes a great deal to the feedback, comments and questions from all my students at the *École nationale supérieure des mines* (engineering school) in Saint Etienne who have "endured" my lecturing in thermodynamics for many years. I am very grateful to them, and also thank them for their stimulating attitude. This work is also the fruit of numerous discussions with colleagues who teach thermodynamics in the largest establishments – particularly in the context of the group "Thermodic", founded by Marc Onillion. My thanks go to all of them for their contributions and conviviality.

This volume is split into two parts. In the first part, Chapter 1 is dedicated to phase modeling tools and covers the modeling of a phase: constructing potential characteristic functions. Chapter 2 covers the microscopic approach and presents the characteristic matrices that group together thermodynamic coefficients, from different experimental data. Chapters 3, 4 and 5 cover tools used in the microscopic modeling of phases through the use of statistics of molecular objects and microcanonical and canonical spaces. The calculation of state functions from molecular data allow the characteristic functions of a phase to be calculated.

The second part covers the modeling of gaseous phases. In Chapter 7, we cover macroscopic and microscopic modeling of pure gases using state equations, generalized compressibility charts and the concept of fugacity. Microscopic modeling is an initial application of statistical thermodynamics by calculating the second virial coefficient. The eighth chapter describes the models of gaseous mixtures, macroscopic and microscopic. Mixed models from models of condensed solutions and state equations are covered.

Michel Soustelle
Saint-Vallier, France
April 2015

Notations

$\{\mathbf{gas}\}$: pure gas

$\{\{\mathbf{gas}\}\}$: gas in mixture

$(\mathbf{liquid})$: pure liquid

$((\mathbf{liquid}))$: liquid in solution

$\langle\mathbf{solid}\rangle$: pure solid

$\langle\langle\mathbf{solid}\rangle\rangle$: solid in solution

Symbols

A : Area of a surface or interface

$A_{\mathrm{H}}^{(12)}$: Hamaker constant between two media 1 and 2

A : affinity

Ã : electrochemical affinity

A_M : molar area

A_m : molecular area

a : cohesion pressure of a gas; radius of the unit cell of a liquid

A, B, ... : components of a mixture

a^{mix} and b^{mix} : terms for mixing of constants of a state equation

B'_i : i^{th} coefficient of the virial in the increase in pressure

B_i : i^{th} coefficient of the virial

b : co-volume of a gas, co-surface of an adsorbed gas

C : concentration, trace concentration of a voltage-pH diagram

C_P^{xs} : excess molar heat capacity at constant pressure

C_i : molar concentration (or molarity) of a component i

$C_\pm$: mean concentration of ions in an ionic solution

$C_{V(el)}$: contribution of free electrons of a metal to the molar heat capacity

$C_{v(r)}$: contribution of rotational movements to the heat capacity at constant volume

$C_{v(t)}$: contribution of translational movements to the heat capacity at constant volume

$C_{v(v)}$: contribution of vibrational movements to the heat capacity at constant volume

C_V, C_P : heat capacity at constant volume and pressure

c : capacity of a capacitor; number of independent components

D : dielectric constant of the medium; protection or contact diameter of a molecule

$D(T/\Theta_D)$: Debye function

d : distance between two liquid molecules

d_eS : exchange of entropy with the external environment

d_i : degree of oxidation i of an element A

d_iS : internal production of entropy

$d\omega$: element of volume

E : energy of the system

Ε : Young's modulus

$E(T/\Theta_E)$: Einstein function

E_0 : internal energy associated with a reaction at temperature 0 K

E^0 : standard electric potential or standard electromotive force of an electrochemical cell

E_{abs} : reversible electrical voltage of an electrochemical cell

Eb : balanced equation

$\langle E \rangle$: total mean energy of an element of a canonical ensemble

E_C : total energy of canonical ensemble

E_I : potential energy due to interactions

E_j : energy of an element j of canonical ensemble

E_{kin} : molar kinetic energy of electrons in a metal

E_p : set of variables at p intensive variables chosen to define a system

e : relative potential of an electrode

e^0 : standard electrical potential (or normal potential) of an electrode

e_0 : potential of equi-activity or equi-concentration of an electrode

e_{abs} : absolute potential of an electrode

F : Helmholtz free energy

F_m^{mix} : excess molar free energy

$\overline{F}_i^{xs}$: excess partial molar Helmholtz free energy of a component i

$\overline{F}_i^{mix}$: partial molar Helmholtz free energy of mixing of a component i

$\overline{F}_i$: Helmholtz free energy, partial molar Helmholtz free energy of a component i

F_{el} : contribution of free electrons to molar Helmholtz free energy

$\tilde{F}$: electrochemical Helmholtz free energy

f_{het} : heterogeneous spreading parameter

f_i : fugacity of component i in a gaseous mixture

F_m : molar Helmholtz free energy

$\mathcal{F}$: faraday

f_i^0 : molar Helmholtz free energy of a pure component i

f^0 or f_i^0 : fugacity of a pure gas i

G_m^{xs} : excess Gibbs free energy

$\tilde{G}_\sigma$: electro-capillary Gibbs free energy

$\tilde{G}$: electrochemical Gibbs free energy

G_i^{xs} : excess partial molar Gibbs free energy of a component i

G : Gibbs free energy

G, $\overline{G}_i$, $[G]$: Gibbs free energy, partial molar Gibbs free energy of i, generalized Gibbs free energy

G_m : molar Gibbs free energy

G_m^{mix} : molar Gibbs free energy of mixing

$\overline{G}_i^{xs}$: excess partial molar Gibbs free energy of a component i

g : osmotic coefficient; acceleration of gravity; coefficients of degeneracy or multiplicity or statistical weight

g_i^0 : molar Gibbs free energy of a pure component i

g_a : statistical weight of fundamental electron level of the nucleus a

g_i : coefficient of multiplicity of state i

$g_{(e)}$: statistic weight of electron levels

g(r) : radial distribution function

$g(v_x)$: distribution of velocity components according to Ox

g^* : molar Gibbs free energy of gas i under a pressure of 1 atmosphere in a mixture

H_T^0 : standard molar enthalpy of formation at temperature T

H, $\overline{H_i}$: enthalpy, partial molar enthalpy of i

H : Hamiltonian

$H_{i,}$: resonance integral between two identical neighboring atoms

$H_{i,i}$: Coulombian integral between two identical neighboring atoms

H: magnetic field

$\tilde{H}$: electrochemical enthalpy

H_m^{xs} : excess molar enthalpy

H_m^{mix} : molar enthalpy of mixing

$\overline{H_i}^{xs}$: excess partial molar enthalpy of a component i

$\overline{H_i}^{mix}$: partial molar enthalpy of mixing of component i

h_t : spreading coefficient

h : stoichiometric coefficient of protons in an electrochemical reaction

h : Planck's constant

h_i^0 : molar enthalpy of a pure component i

$h_{ét}$: spreading coefficient of a liquid on another Harkins liquid

I : ionic strength of an ionic solution

I_m : ionic strength reduced to molarities

I, I_1, I_2 I_3 : moments of inertia.

I_I : configuration integral of the canonical translational partition function

i : Van 't Hoff factor

$\overline{J_i}$: partial molar value of the variable J relative to component i

J_i^{mix} : value of mixing of size J relative to component i

$\overline{J_i^{mix}}$: partial molar value of mixing of size J relative to component i

J_i^* : size of J relative to component i in a perfect solution

$\overline{J_i^*}$: partial molar value of size J relative to component i in a perfect solution

j_i^0 : value of size J for the pure component i in the same segregation state

j : quantum rotation number

$K_{i,j}(E_p)$: Thermodynamic coefficient associated with the set of variables E_p, X_j is its variable of definition and Y_i its function of definition

$K_i^{(Tr)}$: equilibrium constant of the phase transformation Tr for component i

K_{ij} : local composition weighting factor

K_{ads} : adsorption equilibrium constant

K_{AX} : solubility product of solid AX

$K_i^{(\alpha\beta)}$: partition coefficient of compound i between both phases α and β

K_d : dissociation constant

K_{fe} : adsorption equilibrium function

$K_r^{(c)}$: equilibrium constant relative to concentrations

$K_r^{(f)}$: equilibrium constant relative to fugacities

$K_r^{(P)}$: equilibrium constant relative to partial pressures

K_r : equilibrium constant

K_s : solubility product

k : wave number

k_B : Boltzmann constant

L_t : latent heat accompanying the transformation t

l_c : capillary length

M : molar mass

M : magnetic moment; Madelung constant

m , m_s : mass of solute s in grams per kg of solvent

m : total mass

m_i : mass of component i

N : number of molecules in an ensemble

N_a : Avogadro's number

N_A : number of molecules of the component A

N_C : number of elements in the canonic ensemble

N_c : Total number of cells in a liquid.

n : total number of moles in a solution or a mixture; quantum translation number

n_i : number of objects i of system with energy ε_i.

$n^{(\alpha)}$: total number of moles in phase α

$<n>$: mean number of neighboring holes in a molecule in a liquid

n_i : number of moles of component i

N_L : total number of holes in a liquid

P_c^{mix} : critical pressure of a mixture

P : pressure of a gas

P_i^{subl} : sublimation vapor pressure of component i

P_i^{vap}, P_i^0 : saturation vapor pressure of component i

P_r^{mix} : relative pressure of the mixture

P_c : critical pressure

p_F : Fermi impulsion

P_i : partial pressure of component i

P_j : fraction of number of elements in a state j

p : number of external physical variables; spreading parameter

Q : heat balance

Q_a : reaction quotient in terms of activities

Q_P : heat of transformation at constant pressure ; reaction quotient in terms of partial pressure

Q_r : reaction quotient of transformation r

Q_V : heat of transformation at constant volume

q_ϕ : equilibrium heat of adsorption

q_d : differential heat of adsorption.

q_i : volumic fraction parameter

q_{isost} : isosteric heat of adsorption

$\Re$: reaction rate

R : ideal gas constant

R : radius of mean curvature of a surface; distillation reflux ratio

r_A : radius of ionic atmosphere

r_0 : distance of minimal energy between two molecules

r_c : radius of a cylindrical tube

r_i : parameter of volumic fraction

r_K : Kelvin's radius

s : parameter of order of an alloy

S_m^{mix} : molar entropy of mixing

$\overline{S}_i^{xs}$: excess partial molar entropy of component i

$\overline{S}_i^{mix}$: partial molar entropy of mixing of component i

S : oversaturation of a solution

$\overline{S}_i$: entropy, partial molar entropy of i

$\tilde{S}$: electrochemical entropy

S_m^{xs} : excess molar entropy

s_i^0 : molar entropy of the pure component i

T_c^{mix} : critical temperature of the mixture; relative temperature of mixture

T^* : second order transition temperature

$T_{(Az)}$: boiling point of azeotrope

T_c : critical temperature

T_F : Fermi temperature.

$T_i(_{Eb})$: boiling point of pure i

$T_{i(F)}$: melting point of pure i

T_s : sublimation point

T_v : vaporization point

U_m^{xs} : excess molar internal energy

U_m^{mix} : internal molar energy of mixing

$\overline{U}_i^{xs}$: excess partial molar internal energy of component i

$\overline{U}_i^{mix}$: partial molar internal energy of a mixture of component i

U ,$\overline{U}_i$: internal energy, partial molar internal energy of $_i$

U_{el} : contribution of free electrons to molar internal energy

$\tilde{U}$: electrochemical internal energy

U_m : molar internal energy

U_R : internal lattice binding energy

u_+, u- : ionic mobilities of cation and anion

u_i^0 : internal molar energy of pure component i

V, $\overline{V}_i$: volume, partial molar volume of i

V_c : critical volume

V_G : Gibbs variance

V_m : molar volume

v_c : cell volume of a liquid

v_D : Duhem variance

v_f : free volume per molecule

v_∞ : volume of influence around a molecule

v_i^0 : molar volume of the pure component i

v_m : molecular volume

v_M : molar volume of a solid at melting point

v_{mono} : volume of adsorbed gas monolayer

v_{xi} : component according to Ox of the speed of particle i

v : vibrational quantum number

W_{12} : energy per square meter of the interaction between surfaces of phases 1 and 2

w_i : mass fraction of component i

w_{ij} : energy exchanged between atoms i and j

$x_k^{(\alpha)}$: molar fraction of component k in phase α

x, y, z : coordinates of a point in space

x_i : molar fraction of component i in a solution

$<y>$: mean value of y

Y_i and X_i : intensive and extensive conjugate variables

$y_{i,j}$: Mayer's function

y_i : molar fraction of component i in a gaseous phase

Z : compressibility coefficient

Z_i : compressibility coefficient of a gas i

Z^{mix} : compressibility coefficient of a gas mixture

Z_{AB} : molecular partition function of the interaction between molecules

Z_C : canonical partition function

$Z_{C(\text{A})}$: canonical partition function of component A

$Z_{C(I)}$: canonical partition function of interactions

$Z_{C(t)}$: canonical partition function of translation

z : molecular partition function, altitude of a point

z_e : electronic molecular partition function

z_i : electrovalence of ion i; number of nearest neighbor molecules of molecule i

z_{int} : contribution of internal movements to the molecular partition function

z_n : molecular partition function of nucleus

z_{pf} : molecular partition function of a perfect gas

z_r : molecular partition function of rotation

z_t : molecular partition function of translation

$z_{t(pf)}$: molecular partition function of translation of a perfect gas

z_v : molecular partition function of vibration

α: dissociation coefficient of a weak electrolyte; Lagrange multiplier relative to the number of objects in a collection; polarizability of a molecule; linear expansion coefficient under pressure P; relative volatility

α_a : apparent dissociation coefficient of a weak electrolyte

β : Lagrange multiplier relative to the energy of objects in a collection; volumic expansion coefficient under pressure P

$\Gamma(\mathsf{E}_P)$: characteristic function with ensemble E_P as canonical variables

Γ_k : activity coefficient of a group

Γ : characteristic function

Γ_i : excess of surface or surface concentration of component i

$\Gamma_{i,j}$: excess of surface or surface concentration of component i with respect to j

γ : activity coefficient of component i whatever the reference state; Grüneisen parameter; structural coefficient that equals $\sqrt{2}$ for face-centered cubic networks

γ_0 : activity coefficient of a solvent

γ_i : activity coefficient of species i; Grüneisen factor of phonon i

$\gamma_i^{(I)}$: activity coefficient of component i reference pure body

$\gamma_i^{(II)}$: activity coefficient of component i reference solution infinitely dilute

$\gamma_i^{(III)}$: activity coefficient of component i reference molar solution

$\gamma_\pm$: mean activity coefficient of ions in an ionic solution

γ_s : activity coefficient of a solute

$\Delta\sigma$: spreading of a liquid

$\Delta_r A_T^0$: standard value at temperature T of A associated with transformation r

$\Delta_r(A)$: value of A associated with transformation r

$\delta_{i,j}$: Kronecker's symbol

δ : coefficient of increase in pressure at volume V

ε_i^j : Wagner interaction coefficient

ε : electrical permittivity of the medium

ε_0 : electrical permittivity of the vacuum; energy of orientation effect between molecules

ε_{attr} : attraction energy between molecules

ε_c : kinetic energy of a molecule

$\varepsilon_{\text{C-H}}$: energy of C-H bond

ε_d : energy of dispersion effect between molecules

ε_F : Fermi energy

$\varepsilon_{i(e)}$: electronic energy of a molecule i

$\varepsilon_{i(l)}$: interaction energy of a molecule i

$\varepsilon_{i(n)}$: nuclear energy of a molecule i

$\varepsilon_{i(r)}$: rotational energy of a molecule i

$\varepsilon_{i(t)}$: translational energy of a molecule i

$\varepsilon_{i(v)}$: vibrational energy of a molecule $i\varepsilon_{i,j}$: interaction energy between two molecules i and j

ε_{ij} : electron pair energy between atoms i and j

ε_{mm} : switch

ε_p : potential energy of a molecule

ε_{rep} : repulsion energy between molecules

η : viscosity

η_{ij} : Warren and Cowley order parameter

Θ_{D} : Debye vibrational temperature

Θ_{E} : Einstein vibrational temperature

Θ_{r} : characteristic rotational temperature

θ : recovery fraction

θ_1 : fraction on the surface of a component

λ : linear expansion coefficient

λ_{0+}, λ_{0-} : equivalent ionic conductivities of cation and anion

λ_A : absolute activity of component A

λ_i : lateral chemical potential of component i

Λ : equivalent conductivity of an electrolyte; thermal wavelength of a molecule

Λ_0 : equivalent boundary conductivity of an electrolyte

μ_i, $\lfloor\mu_i\rfloor$, $\overline{\mu}_i$: chemical potential of component i or electric dipole moment of molecule i, generalized chemical potential

$\mu_i^{(L)}, \mu_i^{(G)}$: chemical potential of component i liquid or gaseous

$\widetilde{\mu}$: electrochemical potential

ν : vibrational frequency

$\nu_{k(\rho)}$: algebraic stoichiometric number of component A_k in reaction ρ

ν_D : maximal Debye frequency

ν_e : stoichiometric coefficient of electrons in an electrochemical reaction

ξ : reaction extent

Π_d : cut-out pressure

ρ : density of molecules in a spherical crown of radius r; volumic density of electrical charges; volumic mass $\rho(r)$: density of molecules in a chamber

σ : surface energy; symmetry number

σ_e : surface density of electric charges

σ^* : surface tension

τ_+, τ_- : cationic and anionic transport number

Φ : practical osmotic coefficient; expansion pressure

Φ_i : fugacity coefficient of component i in a gaseous mixture

ϕ : coefficient of conductivity of a strong electrolyte; number of phases

φ_i : fugacity coefficient of gas i in a mixture

φ_i and φ_i^0 : fugacity coefficient of a pure gas

ϕ_i : fraction of component volume

χ_i : calorimetric coefficient relative to variable x_i

χ : electric conductivity

χ_T : compressibility coefficient at temperature T

Ψ_i : electrostatic potential of ionic atmosphere

$\Psi(r)$: electrostatic potential

Ψ_{km} : energetic term between two groups

Ψ : wave function

Ω_{BE} : number of complexions in Bose–Einstein statistics

Ω_{C} : number of complexions in Fermi–Dirac statistic

Ω : number of complexions

ω_1 : set of coordinates for position of molecule i

ω_x : rotational velocity component according to Ox

1

Thermodynamic Functions and Variables

In thermodynamics, each state of a system is characterized by the value given by a certain characteristic function, which depends on a certain number of variables: the state variables. The most commonly cited variables include the internal energy U, enthalpy H, Helmholtz energy F and Gibbs energy G. Therefore, the complete properties of a system are known if one of the characteristic functions is known according to the variables chosen to define the problem.

These characteristic functions also possess two properties, according to the two first principles of thermodynamics:

– these are the state functions, that is to say, their variations during a transformation only depend on the initial and final state and their differentials are exact total differentials (see Appendix 3, section A3.2);

– they are each a potential function in a set of given variables, i.e. they have a minimal value for a system in equilibrium and vice versa.

If the system is composed of several phases, the value of the characteristic function of the overall system is the sum of the values of this function for each phase.

NOTE.– We know that if a phase contains several components, the value of the characteristic function of the phase is not the sum of the values of these same functions for each of the components.

1.1. State variables and characteristic functions of a phase

The number and nature of variables, used to study the state or thermodynamic modifications of a phase, are chosen by the operator. This limits the number of variables in order to not complicate the problem.

1.1.1. *Intensive and extensive conjugate variables*

Variables that can be chosen are classed in couples. The product of the variables of the same couple is homogeneous at a given energy, that is to say, the equation with dimensions: ML^2T^{-2}. One of the variables in the couple is an intensive variable, i.e. homogeneous to zero degrees with regards to quantities of matter, and the other is an extensive variable, i.e. homogeneous at degree one with regards to quantities of matter (see Appendix 3, section A3.1).

Table 1.1 shows the main couples, where pressure–volume and temperature–entropy are the two couples most frequently encountered in chemical thermodynamics, with, for each component, the couple quantities of matter–chemical potentials. The couples electric charge–electric potential and area–surface tension are found in electrochemistry and surface chemistry, respectively.

The third column in the table provides an expression of the corresponding work in the forms YdX, X and Y, the extensive variable and intensive variable of the couple, respectively.

Note that with volume being the intensive variable, opposed pressure must respect the form of the pressure work forces.

When dealing with the thermodynamic modeling of a transformation, it is best to first choose the physical phenomena involved. From Table 1.1, the conjugate variable couples used are deduced.

For example, in the case of a steam engine machine in which the phenomena involved are heating and changes in the state of water, the following thermomechanical couples are used: temperature–entropy (T, S) and opposed pressure–volume (-P, V). To study a chemical transformation in condensed phases at atmospheric pressure, incapable of varying or even influencing the phenomenon, the couple temperature–entropy is chosen and, for each component, the couple quantity of matter and chemical potential.

Extensive variable	*Intensive variable*	*Reversible work*
Entropy (S)	Temperature (T)	$T\mathrm{d}S$
Volume (V)	Opposed pressure ($-P$)	$-P\mathrm{d}V$
Area (A)	Surface energy (σ)	$\sigma\mathrm{d}A$
Quantity of matter of i (n_i)	Chemical potential of i (μ_i)	$\mu_i\mathrm{d}n_i$
Electric charge (q)	Electrical potential Φ	$\Phi\mathrm{d}q$
Magnetic moment ($\mathcal{M}$)	Magnetic field ($\mathcal{H}$)	$\mathcal{H}\mathrm{d}\mathcal{M}$
Length (l)	Force (F)	$F\mathrm{d}l$

Table 1.1. *Couples of conjugate variables and corresponding reversible work*

Unlike in thermomechanics, which only involves couples (T, S) and (-P, V), we encounter several chemical thermodynamics. “Simple” chemical thermodynamics involves the couples (T, S) ($-P$, V), associated with each component (μ_i, n_i). “Electrochemical” thermodynamics implicates electrical phenomena and therefore involve the following couples: (T, S) ($-P$, V), (μ_i, n_i) and (Φ, q). “Capillary” chemical thermodynamics implicates surface phenomena and involves the couples (T, S) ($-P$, V), (μ_i, n_i) and (σ, A). Finally, “electrocapillary” chemical thermodynamics involves the couples (T, S) ($-P$, V), (μ_i, n_i), (Φ, q) and (σ, A).

1.1.2. *Variations in internal energy during a transformation*

Let us imagine that the set $\mathcal{E}_0$ is defined as the total of n extensive variables chosen for the phenomenon studied.

$$\mathcal{E}_0 = (X_1, X_2, \ldots, X_i, \ldots X_n) \qquad [1.1]$$

We know that the variation in internal energy for an infinite transformation is written:

$$\mathrm{d}U = \delta W + \delta Q = \sum_{i=1}^{n} Y_i \,\mathrm{d} X_i \qquad [1.2]$$

Now the internal energy is a state function and therefore the previous differential is an exact total differential (see Appendix 3, section A3.2) and will be known when all the partial derivatives attached to ensemble $\mathcal{E}_0$ are known, such that:

$$\left(\frac{\partial U}{\partial X_i}\right)_{\mathcal{E}_0} = Y_i(X_1, X_2, \ldots X_i, \ldots X_n) \qquad [1.3]$$

For example, for a thermomechanical transformation of a pure body, the set $\mathcal{E}_0$ will be defined as:

$$\mathcal{E}_0 = (S, V)$$

The variation in internal energy will be:

$$dU = \delta W + \delta Q = T\,\mathrm{d}S - P\,\mathrm{d}V$$

And the first derivatives of U will be:

$$\left(\frac{\partial U}{\partial S}\right)_V = T(S,V) \quad \text{and} \quad \left(\frac{\partial U}{\partial V}\right)_S = -P(S,V)$$

However, choosing the set $\mathcal{E}_0$ may appear to be poorly suited to the problem, for example, if we wish to calculate the variation in internal energy in an infinitesimal transformation, which may be done by choosing the set of variables, $\mathcal{E}_p$ is instead defined by:

$$\mathcal{E}_p = (Y_1, Y_2, \ldots, Y_i, \ldots Y_p, X_{p+1}, \ldots, X_k, \ldots X_n) \qquad [1.4]$$

In this set we chose the following as independent variables:

– the intensive variables for p couples;

– and the extensive variables for the n-p remaining couples.

The variation in internal energy is always written:

$$\mathrm{d}U = \delta W + \delta Q = \sum_{i=1}^{n} Y_i \,\mathrm{d}X_i = \sum_{i=1}^{p} Y_i \,\mathrm{d}X_i + \sum_{k=p+1}^{n} Y_k \,\mathrm{d}X_k \qquad [1.5]$$

We can always express the parameters X_i according to the variables chosen and state the expressions in relation to [1.5]. This rapidly results in complicated expressions of dU.

We show this change in variables with both thermomechanical couples by choosing the set:

$$\mathcal{E}_1 = (T, V)$$

The differential of entropy is always expressed according to new variables:

$$\mathrm{d}S = \left(\frac{\partial S}{\partial T}\right)_V \mathrm{d}T + \left(\frac{\partial S}{\partial V}\right)_T \mathrm{d}V$$

By writing in terms of dU, the differential of internal energy appears in the new set of variables:

$$\mathrm{d}U = T\,\mathrm{d}S - P\,\mathrm{d}V = T\left(\frac{\partial S}{\partial T}\right)_V \mathrm{d}T + \left[\left(\frac{\partial S}{\partial V}\right) - P\right]\mathrm{d}V$$

1.1.3. *Characteristic function associated with a canonical set of variables*

We have seen that the method, involving changing the variables to express the internal energy, quickly results in complex expressions, which is why we prefer to use a better adapted characteristic function Γ for each variable set, which has the same properties as the internal energy, i.e. it remains a characteristic function whose knowledge will allow, as for U in $\mathcal{E}_0$ variables, a complete definition of the phase studied in the thermodynamic plane, and this remains a state function. However, this function will be expressed in a set of variables to which it will be adapted, called its canonical variables.

Let us take our set of variables $\mathscr{E}_0$ defined by the relation [1.4] and the expression of internal energy according to relation [1.5]. We define the characteristic function Γ_p, with canonical variables in set $\mathscr{E}_p$, by the following transformation, which is called a Legendre transformation:

$$\Gamma(\mathscr{E}_p) = U - \sum_{i=1}^{p} X_i Y_i \qquad [1.6]$$

This transformation involves subtracting from the internal energy all the products $X_i Y_i$ for which the intensive Y_i was chosen as the problem variable. $\Gamma(\mathscr{E}_p)$ is a state function and by differentiating the expression [1.9] we obtain:

$$\mathrm{d}\Gamma(\mathscr{E}_p) = \sum_{i=1}^{p} X_i \,\mathrm{d}Y_i + \sum_{k=p+1}^{n} Y_k \,\mathrm{d}X_k \qquad [1.7]$$

We see that the variations in the function chosen are expressed according to the variables chosen. This variation will be fully known if the following partial derivatives are also known:

$$\left(\frac{\partial \Gamma(\mathscr{E}_p)}{\partial Y_i}\right)_{\mathscr{E}_p} = X_i \quad \text{and} \quad \left(\frac{\partial \Gamma(\mathscr{E}_p)}{\partial X_k}\right)_{\mathscr{E}_p} = Y_k$$

Which are written in the general form:

$$\left(\frac{\partial \Gamma(\mathscr{E}_p)}{\partial Z_m}\right)_{\mathscr{E}_p} = \varepsilon_{mm} \Psi_m \qquad [1.8]$$

Ψ_m is the conjugate of variable Z_m. ε_{mm} is a switch which takes the value:

- $\varepsilon_{mm} = +1$ if variable Z_m is extensive (Ψ_m is therefore intensive);
- $\varepsilon_{mm} = -1$ if variable Z_m is intensive (Ψ_m is therefore extensive).

Certain frequently encountered characteristic functions have a particular name; these are displayed in Table 1.2.

Generalized Gibbs energy, denoted $[G]$ is the characteristic function obtained by choosing the amounts of matter and all other intensive variables of the problem as variables.

$$[G] = U - \sum_{i=2}^{p} X_i Y_i \qquad [1.9]$$

This function is confused with Gibbs energy if, as well as the amount of matter, only the thermomechanical couples S, T and V, $-P$ are considered. In the case where an electrical potential is added, the function is called electrochemical Gibbs energy and its variables are temperature, pressure, electrical potential and amount of matter. For surface phenomena, we encounter capillary Gibbs energy which involves the following variables: temperature, pressure, surface tension and amount of matter.

Variables chosen	*Characteristic function*
Extensive variables	U Internal energy
Extensive variables except V replaced by $-P$	H Enthalpy
Extensive variables except S replaced by T	F Helmholtz energy
V and S replaced by $-P$ and T	G Gibbs energy

Table 1.2. *Characteristic functions and associated canonical variables*

1.2. Partial molar parameters

The concept of partial molar parameter stems from phases with several components; extensive parameters are not additive, i.e. a given extensive variable of the phase, for example volume, is not the sum of extensive variables of different components. The partial molar parameter of a component presents itself as the contribution of one mole of this compound to the value of the corresponding extensive parameter of the phase.

1.2.1. *Definition*

In this section we will consider phases with N components and we will choose the following variables; the N quantities of matter n_i and the p intensive variables for the other conjugate couples. We call the partial molar variable of an extensive function its derivative with respect to the amount of matter of a compound whose intensive variables and all the amounts of matter of the other components are constant. For the extensive variable Z, this partial molar variable is denoted $\overline{Z_i}$. We therefore have, by definition:

$$\overline{Z_i} = \left(\frac{\partial Z}{\partial n_i}\right)_{Y_k, n_j} \quad [1.10]$$

1.2.2. *Properties of partial molar variables*

Let us recap the following relations which are found in all basic publications of chemical thermodynamics:

$$Z = \sum_{i=1}^{N} n_i \overline{Z_i} \quad [1.11]$$

$$\sum_i n_i \, \mathrm{d}\overline{Z_i} = \sum_{k=1}^{p} \left(\frac{\partial Z}{\partial Y_k}\right) \mathrm{d}Y_k \quad [1.12]$$

This last relation is known as the generalized Gibbs–Duhem relation. If we keep constant all the intensive variables Y_k, this relation can be simplified to:

$$\sum_i n_i \, \mathrm{d}\overline{Z_i} = 0 \quad [1.13]$$

We can then use this relation using the molar fractions x_i:

$$\sum_i x_i \, \mathrm{d}\overline{Z_i} = 0 \quad [1.14]$$

1.3. Chemical potential and generalized chemical potentials

We are going to assess some properties of the chemical potential and how it can be extended to generalized chemical potentials.

1.3.1. *Chemical potential and partial molar free enthalpy*

We are going to prove the following theorem:

THEOREM 1.– The chemical potential is confused with partial molar Gibbs energy.

Let us consider the study of a phase with N components in which we consider the couples of variables (T, S), ($-P$, V), the chemical potentials and amounts of matter.

The variation in internal energy of our phase is written:

$$dU = T\,dS - P\,dV + \sum_{i=1}^{N} \mu_i\,dn_i \qquad [1.15a]$$

The variation in Gibbs energy:

$$dG = -S\,dT + V\,dP + \sum_{i=1}^{N} \mu_i\,dn_i \qquad [1.15b]$$

In addition, we can directly write the Gibbs energy as an exact total differential (see Appendix 3, section A3.2):

$$dG = \frac{\partial G}{\partial T}dT + \frac{\partial G}{\partial P}dP + \sum_{i=1}^{N}\left(\frac{\partial G}{\partial n_i}\right)_{P,T,n_j} dn_i \qquad [1.16]$$

We therefore obtain:

$$\mu_i = \left(\frac{\partial G}{\partial n_i}\right)_{P,T,n_j} = \overline{G_i} \qquad [1.17]$$

This verifies our theorem.

1.3.2. *Definition of generalized chemical potential*

We call the generalized chemical potential the partial molar Gibbs energy (see section 1.2 and relation [1.10]) i.e.:

$$[\mu_i] = \left(\frac{\partial G_G}{\partial n_i}\right)_{Y,n_{ij}} = [G_i] \qquad [1.18]$$

1.3.3. *Variations in the chemical potential and generalized chemical potential with variables*

1.3.3.1. *Variations with intensive variables*

Later we will see that through the simple application of the symmetry property of the characteristic matrix (see section 2.1.4), we immediately obtain:

$$\frac{\partial[\mu_i]}{\partial Y_k} = -\overline{X_{k_i}} \qquad [1.19]$$

In particular, for conventional phases with several components simply with the couples (S, T) and (P, V), in its amounts of matter, we obtain:

$$\frac{\partial \mu_i}{\partial T} = -\overline{S_i} \quad \text{and} \quad \frac{\partial \mu_i}{\partial P} = \overline{V_i}$$

1.3.3.2. *Variations with composition*

The symmetry properties of the conventional matrix also give:

$$\frac{\partial[\mu_i]}{\partial n_j} = \frac{\partial[\mu_j]}{\partial n_i} \qquad [1.20]$$

1.3.3.3. *Total differential*

By grouping the expressions [1.19] and [1.20], we obtain:

$$\mathrm{d}[\mu_i] = -\sum_k \overline{X_{k_i}}\,\mathrm{d}Y_k + \sum_{j=1}^{N} \frac{\partial[\mu_i]}{\partial n_j}\mathrm{d}n_j \qquad [1.21]$$

1.3.4. *Gibbs–Duhem relation*

Since the chemical potential is a partial molar variable, we can apply the relations [1.11], [1.12] and [1.13] and write:

$$[G] = \sum_{i=1}^{N} [\mu_i] n_i \qquad [1.22]$$

The Gibbs–Duhem law can be written in its generalized form:

$$\sum_{i=1}^{N} n_i \, \mathrm{d}[\mu_i]_i - \sum_{k=1}^{p} X_k \, \mathrm{d} Y_k = 0 \qquad [1.23]$$

and in its simplified form, with physical intensive variables kept constant:

$$\sum_{i=1}^{N} n_i \, \mathrm{d}[\mu_i] = 0 \qquad [1.24]$$

By taking into account the equality [1.21], this relation above gives:

$$\sum_{j=1}^{N} \frac{\partial [\mu_i]}{\partial n_j} \mathrm{d} n_j = 0 \qquad [1.25]$$

1.3.5. *Generalized Helmholtz relations*

Let us calculate the expression $\dfrac{\partial([G]/T)}{\partial T}$:

$$\frac{\partial([G]/T)}{\partial T} = \frac{1}{T}\frac{\partial([G])}{\partial T} - \frac{[G]}{T^2} = \frac{S}{T} - \frac{[G]_G}{T^2} \qquad [1.26]$$

Let us write:

$$\Gamma(\xi_{p\ 1}) = [G] + TS \qquad [1.27]$$

$\Gamma(\mathcal{E}_{p-1})$ is the characteristic function associated with the set of variables:

– the amounts of matter;

– entropy;

– all the intensive variables other than temperature and chemical potentials.

By comparison, we deduce the first Helmholtz relation:

$$\frac{\partial([G]/T)}{\partial T} = -\frac{\Gamma(\mathcal{E}_{p-1})}{T^2} \quad [1.28]$$

For example, in the particular case of conventional chemical phases (variables P and T), it is written:

$$\frac{\partial(G/T)}{\partial T} = -\frac{H}{T^2} \quad [1.29]$$

H is the enthalpy.

By deriving the expression [1.28] relative to the amount n_i of component i, we obtain the second Helmholtz relation:

$$\frac{\partial(\overline{G}/T)}{\partial T} = -\frac{\overline{\Gamma(\mathcal{E}_{p-1})}_i}{T^2} \quad [1.30]$$

$\overline{\Gamma(\mathcal{E}_{p-1})}_i$ is the partial molar variable relative to $\Gamma(\mathcal{E}_{p-1})$.

With regard to conventional chemical phases, this relation is written:

$$\frac{\partial(\mu_i/T)}{\partial T} = -\frac{\overline{H}_i}{T^2} \quad [1.31]$$

$\overline{H}_i$ is the partial molar enthalpy of component i.

1.3.6. *Chemical system associated with the general system*

If we consider the phase requiring the variable couples (T, S), (P, T) and (Y_k, X_k), we call the chemical system associated with this phase the identical phase in which the variables Y_k and X_k are not implicated, or in which one of the two variables of couples X_k, Y_k (chosen as the variable of the problem) is negligible.

For example, a phase defined from the set of variables: P, T, and Φ, is an electrochemical system. The chemical phase associated would be the same phase under zero potential (or zero charge). The generalized Gibbs energy of this phase is the electrochemical Gibbs energy, denoted $\widetilde{G}$ and defined by:

$$\widetilde{G} = U + PV - TS - q\Phi = \widetilde{U} + PV - T\widetilde{S} \qquad [1.32]$$

We define the electrochemical potential $\widetilde{\mu}_i$ as the partial molar variable of $\widetilde{G}$ for component i.

Let us consider the state function as defined by:

$$U_{chem} = \widetilde{U} - q\Phi \qquad [1.33]$$

U_{chem} is the internal energy of the phase of the chemical system associated with our electrochemical phase.

By differentiating under zero potential, we obtain:

$$\mathrm{d}U_{chem} = T\,\mathrm{d}S - P\,\mathrm{d}V + \sum \mu_i\,\mathrm{d}n_i \qquad [1.34]$$

μ_i is the chemical potential of i in the chemical system associated with our electrochemical phase. We can write:

$$\widetilde{\mu}_i = \mu_i + q\Phi \qquad [1.35]$$

However, if $\mathfrak{F}$ designates the Faraday constant and z_i the electrovalence of component i, the charge is:

$$q = \sum_i n_i z_i \mathfrak{F} \qquad [1.36]$$

By referring to expression [1.35], the electrochemical potential becomes:

$$\tilde{\mu}_i = \mu_i + \mathfrak{F}\Phi\sum_i n_i z_i \qquad [1.37]$$

The same for the variables mass and altitude, we define a gravitational chemical potential according to:

$$[\mu_i] = \mu_i + g\sum_i n_i M_i \qquad [1.38]$$

g is the acceleration due to gravity and M_i the molar mass of component i.

1.4. The two modeling scales

With the characteristic functions not being directly accessible from the experiment, there are two possible methods to calculate them: the first method, called macroscopic modeling, is based on the macroscopic variables; the second method, called microscopic modeling, is based on molecular data and statistical units.

In macroscopic modeling, these functions are obtained from a certain amount of data including state equations, thermomechanical coefficients, heat capacities, variations in surface tensions, Young's modulus and all variables which are classed under thermodynamic coefficients, themselves grouped into characteristic matrices. This will be the focus of Chapter 2, in which we introduce these concepts and study this method.

Microscopic modeling is based on the statistics of collections of discernible or indiscernible objects, the introduction of canonical and molecular partition functions and the use of data about the molecule obtained from spectroscopy. Chapters 3, 4, 5 and 6 will develop these concepts and follow onto microscopic modeling.

2

Macroscopic Modeling of a Phase

To form potential functions in macroscopic modeling, one must be able to calculate the characteristic function of a phase from experimental data. Here the challenge is to determine which experimental data must be used for a given phase with a selected ensemble of variables. To solve this, we will introduce the concepts of characteristic matrix and thermodynamic coefficient.

2.1. Thermodynamic coefficients and characteristic matrices

We will associate with each characteristic function, and thus each ensemble of variables, a characteristic matrix formed from the associated thermodynamic coefficients. A certain number of matrix properties reduced the number of terms needed to define a thermodynamic state. These terms will therefore be the only ones required to solve any problem regarding the transformation of the studied phase.

2.1.1. *Thermodynamic coefficients and characteristic matrix associated with the internal energy*

Let us consider a phase for which we have selected, as variables, those in ensemble $\mathcal{E}_0$, as defined in the previous chapter (Equation [1.1]) by the ensemble of relevant extensive variables. The state of this phase and its reversible evolution will be known if the internal energy U and its variations as a function of the extensive variables of ensemble $\mathcal{E}_0$ are fully known.

According to relation [1.2], we find that knowing only the intensive conjugate variables as a function of the variables of ensemble $\mathcal{E}_0$ is sufficient:

$$Y_i = f_i\left(X_1, X_2, ..., X_i, ...X_n\right) \qquad [2.1]$$

Similarly, to know a particular value for each variable Y_i and the terms of the squared matrix formed on the n^2 partial derivatives:

$$K_{i,j}(\mathcal{E}_0) = \left(\frac{\partial Y_i}{\partial X_j}\right)_{\mathcal{E}_0} \qquad [2.2]$$

We call this matrix $\mathcal{M}(\mathcal{E}_0)$ *the matrix associated with the internal energy* or the ensemble of canonical variables of *U,* i.e. the ensemble $\mathcal{E}_0$ of relevant extensive variables for the phase.

The terms $K_{i,j}(\mathcal{E}_0)$ defined by relation [2.2] are called the *thermodynamic coefficients associated with the ensemble of variables* $\mathcal{E}_0$. X_j is the definition variable of the thermodynamic coefficient and Y_i is the definition function.

The characteristic matrix [2.3] will be formed as follows:

– each line is characterized by a definition variable;

– each column is characterized by a definition function.

The first diagonal of the matrix is composed of thermodynamic coefficients whose definition function and definition variable are conjugate variables. The matrix will be written as follows:

$$\mathcal{M}(\mathcal{E}_0) = \begin{Vmatrix} \frac{\partial Y_1}{\partial X_1} \frac{\partial Y_2}{\partial X_1} \cdots \frac{\partial Y_i}{\partial X_1} \cdots \frac{\partial Y_n}{\partial X_1} \\ \frac{\partial Y_1}{\partial X_2} \frac{\partial Y_2}{\partial X_2} \cdots \frac{\partial Y_i}{\partial X_2} \cdots \frac{\partial Y_n}{\partial X_2} \\ ------ \\ \frac{\partial Y_1}{\partial X_j} \frac{\partial Y_2}{\partial X_j} \cdots \frac{\partial Y_i}{\partial X_j} \cdots \frac{\partial Y_n}{\partial X_j} \\ ------ \\ \frac{\partial Y_1}{\partial X_n} \frac{\partial Y_2}{\partial X_n} \cdots \frac{\partial Y_i}{\partial X_n} \cdots \frac{\partial Y_n}{\partial X_n} \end{Vmatrix} \qquad [2.3]$$

2.1.2. *Symmetry of the characteristic matrix*

It is easy to prove the following theorem:

THEOREM 1.– The thermodynamic coefficients are the secondary partial derivatives of the internal energy.

As these partial derivatives do not depend on the order of derivation, we deduce the following theorem:

THEOREM 2.– The characteristic matrix $\mathcal{M}(\mathcal{E}_0)$ is symmetrical.

Where the ensemble of relations:

$$K_{i,j}(\mathcal{E}_0) = K_{j,i}(\mathcal{E}_0) \qquad [2.4]$$

are the Maxwell relations for the internal energy.

2.1.3. *The thermodynamic coefficients needed and required to thermodynamically define the phase*

According to the previous theorem, to determine the phase on the thermodynamic plane, the following must be known:

– the terms of the main diagonal of the matrix;

– either one of two symmetrical terms $K_{i,j}(\mathcal{E}_0)$ or $K_{j,i}(\mathcal{E}_0)$.

Finally let $n(n+1)/2$ be the thermodynamic coefficients instead of the n^2 terms of the matrix.

These coefficients are clearly functions of the selected $\mathcal{E}_0$ variables, but these variations are not undefined; they must satisfy the Schwartz relations which can be written as follows:

$$\frac{\partial K_{i,j}(\mathcal{E}_0)}{\partial X_k} = \frac{\partial K_{i,k}(\mathcal{E}_0)}{\partial X_j} \qquad [2.5]$$

So by taking these equalities into account, we find that to determine the phase, we must know:

– the variations in the coefficients of the main diagonal, each as a function of the definition variable:

$$K_{i,i}(\mathscr{E}_0) = g(X_i) \quad [2.6]$$

– the variations in the other coefficients needed (one of two symmetrical terms) as a function of:

- their definition variable;

- the conjugate variable of their definition function;

- the definition variables of the coefficients located in the same matrix in the same column and higher than the coefficient considered. This is reflected in the functions:

$$K_{i,j}(\mathscr{E}_0) = h\ (X_i, X_j, X_{j-1}, X_{j-2}, \ldots, X_1) \quad [2.7]$$

So, for the thermodynamic example with the ensemble of variables:

$$\mathscr{E}_0 = (S,V)$$

The characteristic matrix is:

$$\mathscr{M}(\mathscr{E}_0) = \begin{Vmatrix} \left(\dfrac{\partial T}{\partial S}\right)_V & \left(-\dfrac{\partial P}{\partial S}\right)_V \\ \left(\dfrac{\partial T}{\partial V}\right)_S & \left(-\dfrac{\partial P}{\partial V}\right)_S \end{Vmatrix}$$

To determine the phase, the following functions must be known:

$$\left(\frac{\partial T}{\partial S}\right)_V = f_1(S)\ ;\ \left(-\frac{\partial P}{\partial V}\right)_S = f_2(V)$$

and

$$\left(\frac{\partial T}{\partial V}\right)_S = f_3(S,V) \text{ or } \left(-\frac{\partial P}{\partial S}\right)_S = f_4(S,V)$$

since the symmetry gives:

$$\left(\frac{\partial T}{\partial V}\right)_S = -\left(\frac{\partial P}{\partial S}\right)_V$$

2.1.4. *Choosing other variables: thermodynamic coefficients and characteristic matrix associated with a characteristic function*

We have seen that the ensemble of variables $\mathcal{E}_0$ is not always best adapted to the phenomenon descriptions. Now, let us consider another ensemble of variables and ensemble $\mathcal{E}_p$ defined by relation [1.4] which is obtained from $\mathcal{E}_0$ by exchanging the extensive variables X_1, X_2, ... X_p with the corresponding conjugate intensive variables. $\mathcal{E}_p$ is therefore defined by:

$$\mathcal{E}_p = (Y_1, Y_2, ..., Y_i, ...Y_p, X_{p+1}, ..., X_k, ...X_n)$$

The corresponding characteristic function is defined by Equation [1.6] and its differential is given by Equation [1.7].

The matrix of characteristic coefficients is defined by: $\left\|K_{i,j}(\mathcal{E}_p)\right\|$, whose terms have one of the four following forms:

$$K_{k,i}(\mathcal{E}_p) = \left(\frac{\partial Y_k}{\partial Y_i}\right)_{\mathcal{E}_p} \qquad K_{i,k}(\mathcal{E}_p) = \left(\frac{\partial X_i}{\partial X_k}\right)_{\mathcal{E}_p} \qquad [2.8]$$

$$K_{i,j}(\mathcal{E}_p) = \left(\frac{\partial X_i}{\partial Y_j}\right)_{\mathcal{E}_p} \qquad K_{k,m}(\mathcal{E}_p) = \left(\frac{\partial Y_k}{\partial X_m}\right)_{\mathcal{E}_p} \qquad [2.9]$$

The characteristic matrix will be defined by the following matrix product:

$$\mathcal{M}(\mathcal{E}_p) = \left\|\text{diag}.\varepsilon_{jj}(\mathcal{E}_p)\right\|\left\|K_{i,j}(\mathcal{E}_p)\right\| \qquad [2.10]$$

The terms $\varepsilon_{jj}(\mathcal{E}_p)$ of the main diagonal matrix will be equal to -1 with j somewhere between 1 and p, when the chosen variable is intensive, and equal to +1 if j is between p+1 and n, i.e. if the chosen variable is extensive.

It is easy to prove, as in section 2.1.2 in the case of internal energy, the two following theorems:

THEOREM 3.– The terms of the characteristic matrix are the secondary derivatives of the characteristic function.

NOTE.– The previous terms are no longer thermodynamic coefficients because of terms $\varepsilon_{jj}(\mathcal{E}_p)$.

THEOREM 4.– The characteristic matrix is still symmetrical (which is no longer the case for the matrix of the thermodynamic coefficients).

The definition [2.10] includes the particular case of ensemble $\mathcal{E}_0$ mentioned in section 2.1.3, with $\varepsilon_{jj}(\mathcal{E}_0) = +1$ for j.

So, to completely determine the phase in the thermodynamic plane, we must determine the following in the characteristic matrix, in the same way as for M_0:

– the variations in coefficients of the main diagonal as a function of their only definition variable;

– the variations in the other coefficients needed (one of two symmetrical terms) as a function of:

- their definition variable;

- the conjugate variable of their definition function;

- the definition variables of the coefficients situated in the matrix in the same column and higher than the coefficient considered.

To illustrate this, let us take the example of a phase with two components, 1 and 2, with the following ensemble of variables:

$$\mathcal{E}_2 = (-P, T, n_1, n_2)$$

in which there are two intensive variables (T and $-P$) and two extensive variables: the amounts of matter.

The corresponding characteristic function is the Gibbs free energy defined by:

$$G = U + PV - TS$$

The matrix of the thermodynamic coefficients is:

$$\left\|K_{i,j}(\mathcal{E}_2)\right\| = \begin{Vmatrix} \left(\frac{\partial S}{\partial T}\right) & \left(\frac{\partial V}{\partial T}\right) & \left(\frac{\partial \mu_1}{\partial T}\right) & \left(\frac{\partial \mu_2}{\partial T}\right) \\ \left(-\frac{\partial S}{\partial P}\right) & \left(-\frac{\partial V}{\partial P}\right) & \left(-\frac{\partial \mu_1}{\partial P}\right) & \left(-\frac{\partial \mu_2}{\partial P}\right) \\ \left(\frac{\partial S}{\partial n_1}\right) & \left(\frac{\partial V}{\partial n_1}\right) & \left(\frac{\partial \mu_1}{\partial n_1}\right) & \left(\frac{\partial \mu_2}{\partial n_1}\right) \\ \left(\frac{\partial S}{\partial n_2}\right) & \left(\frac{\partial V}{\partial n_2}\right) & \left(\frac{\partial \mu_1}{\partial n_2}\right) & \left(\frac{\partial \mu_2}{\partial n_2}\right) \end{Vmatrix}$$

The characteristic matrix is calculated from:

$$\mathcal{M}(\mathcal{E}_2) = \begin{Vmatrix} -1 & 0 & 0 & 0 \\ 0 & -1 & 0 & 0 \\ 0 & 0 & 1 & 0 \\ 0 & 0 & 0 & 1 \end{Vmatrix} \left\|K_{i,j}(\mathcal{E}_2)\right\|$$

Therefore, the characteristic matrix becomes:

$$\mathcal{M}(\mathcal{E}_2) = \begin{Vmatrix} \left(-\frac{\partial S}{\partial T}\right) & \left(-\frac{\partial V}{\partial T}\right) & \left(-\frac{\partial \mu_1}{\partial T}\right) & \left(-\frac{\partial \mu_2}{\partial T}\right) \\ \left(\frac{\partial S}{\partial P}\right) & \left(\frac{\partial V}{\partial P}\right) & \left(\frac{\partial \mu_1}{\partial P}\right) & \left(\frac{\partial \mu_2}{\partial P}\right) \\ \left(\frac{\partial S}{\partial n_1}\right) & \left(\frac{\partial V}{\partial n_1}\right) & \left(\frac{\partial \mu_1}{\partial n_1}\right) & \left(\frac{\partial \mu_2}{\partial n_1}\right) \\ \left(\frac{\partial S}{\partial n_2}\right) & \left(\frac{\partial V}{\partial n_2}\right) & \left(\frac{\partial \mu_1}{\partial n_2}\right) & \left(\frac{\partial \mu_2}{\partial n_2}\right) \end{Vmatrix}$$

So, to define the phase in the thermodynamic plane, we must know:

– the four coefficients of the main diagonal as a function of their definition variable:

$$\left(\frac{\partial S}{\partial T}\right)(T),\ \left(\frac{\partial V}{\partial P}\right)(P),\ \left(\frac{\partial \mu_1}{\partial n_1}\right)(n_1),\ \left(\frac{\partial \mu_2}{\partial n_2}\right)(n_2)$$

– one of two symmetrical coefficients according to the suitable variables, for example:

$$\left(\frac{\partial V}{\partial T}\right)(P,T),\ \left(\frac{\partial \mu_1}{\partial P}\right)(P,T,n_1),\ \left(\frac{\partial \mu_2}{\partial P}\right)(P,T,n_2)\ ;\ \left(\frac{\partial \mu_1}{\partial T}\right)(T,n_1)$$

$$\left(\frac{\partial \mu_2}{\partial T}\right)(T,n_2)\ ;\ \left(\frac{\partial \mu_2}{\partial n_{1,}}\right)(P,T,n_{1,}n_2)$$

NOTE.– By convention, we always put in the top left (see matrix $\mathfrak{M}(\mathcal{E}_2)$ above) the coefficients associated with couples (S,T) and ($V,-P$), which always give the base variables for all other coefficients. In fact, it is easier to obtain the function $\partial q/\partial E = f(T)$ which, as we will see, reflects the variation in electric permittivity according to temperature, than function $\partial q/\partial T = f(E)$.

2.1.5. *Change in variable from one characteristic matrix to another*

Now, if we know the thermodynamic coefficients relative to an ensemble of variables, we can calculate all the other thermodynamic coefficients relative to any other variables, belonging to the same couples of conjugate variables as above.

We will calculate the passage from the ensemble of variables $\mathcal{E}_0$ defined by the ensemble of extensive variables:

$$\mathcal{E}_0 = (X_1, X_2, ..., X_i, ..., X_n)$$

to the variables $\mathcal{E}_{1,}$ which only differ from the previous by the exchange of variables X_l and Y_l:

$$\mathcal{E}_1 = (Y_1, X_2, ..., X_i, ..., X_n)$$

Any Y_i variable for $i \neq 1$ is a function in both ensembles of variables and therefore dY_i can be expressed in two different ways:

– in ensemble $\mathcal{E}_1$:

$$\mathrm{d}Y_i = K_{i,1}(\mathcal{E}_1)\,\mathrm{d}Y_1 + \sum_{j=2}^{n} K_{i,j}(\mathcal{E}_1)\,\mathrm{d}X_j \qquad [2.11]$$

– and for dY1, in ensemble $\mathcal{E}_0$:

$$\mathrm{d}Y_1 = \sum_{j=1}^{n} K_{1,j}(\mathcal{E}_0)\,\mathrm{d}X_j \qquad [2.12]$$

By inserting this differential in the first term of the second part of relation [2.11] and by identifying the coefficients of dX_j, we obtain:

– for $i \neq 1$ and $j = 1$:

$$K_{i,1}(\mathcal{E}_1) = \frac{K_{i,1}(\mathcal{E}_0)}{K_{1,1}(\mathcal{E}_0)} = -K_{1,i}(\mathcal{E}_1) \qquad [2.13]$$

– for $i \neq 1$ and $j > 1$:

$$K_{i.j}(\mathcal{E}_1) = K_{i.j}(\mathcal{E}_0) = -K_{1.j}(\mathcal{E}_0)\frac{K_{i.1}(\mathcal{E}_0)}{K_{1.1}(\mathcal{E}_0)} \qquad [2.14]$$

– and, of course:

$$K_{1,1}(\mathcal{E}_1) = \frac{1}{K_{1,1}(\mathcal{E}_0)} \qquad [2.15]$$

The three relations above allow the thermodynamic coefficients relative to the ensemble of variables $\mathcal{E}_1$ to be calculated, knowing these relative to the ensemble of variables $\mathcal{E}_0$. Systematically, we can therefore obtain similar relations to move from $\mathcal{E}_1$ to $\mathcal{E}_2$, from $\mathcal{E}_2$ to $\mathcal{E}_3$ etc. So to move from $\mathcal{E}_p$ to $\mathcal{E}_{p+1}$, we obtain the relations:

– for $i \neq p$:

$$K_{i,p}(\mathcal{E}_p) - \frac{K_{i,p}(\mathcal{E}_{p-1})}{K_{p,p}(\mathcal{E}_{p-1})} = \frac{\varepsilon_{i,i}(\mathcal{E}_p)}{\varepsilon_{p,p}(\mathcal{E}_p)} K_{p,i}(\mathcal{E}_p) \qquad [2.16]$$

– for $i \neq 1$ and $j \neq p$:

$$K_{i,j}(\mathcal{E}_p) = K_{i,j}(\mathcal{E}_{p-1}) = -K_{p,j}(\mathcal{E}_{p-1})\frac{K_{i,p}(\mathcal{E}_{p-1})}{K_{p,p}(\mathcal{E}_{p-1})} \quad [2.17]$$

– and:

$$K_{p,p}(\mathcal{E}_p) = \frac{1}{K_{p,p}(\mathcal{E}_{p-1})} \quad [2.18]$$

Similarly, in the opposite direction, to move from $\mathcal{E}_p$ to $\mathcal{E}_{p-1}$, we obtain:

– for $i \neq p$:

$$K_{i,p}(\mathcal{E}_{p-1}) = \frac{K_{i,p}(\mathcal{E}_{p-1})}{K_{p,p}(\mathcal{E}_p)} = \frac{\varepsilon_{i,i}(\mathcal{E}_{p-1})}{\varepsilon_{p,p}(\mathcal{E}_{p-1})}K_{p,i}(\mathcal{E}_{p-1}) \quad [2.19]$$

– for $i \neq 1$ and $j \neq p$:

$$K_{i,j}(\mathcal{E}_{p-1}) = K_{i,j}(\mathcal{E}_p) = -K_{p,j}(\mathcal{E}_p)\frac{\varepsilon_{p,p}(\mathcal{E}_{p-1})}{\varepsilon_{j,j}(\mathcal{E}_{p-1})}\frac{K_{i,p}(\mathcal{E}_p)}{K_{p,p}(\mathcal{E}_p)} \quad [2.20]$$

– and:

$$K_{p,p}(\mathcal{E}_p) = \frac{1}{K_{p,p}(\mathcal{E}_{p-1})} \quad [2.21]$$

We can therefore obtain all the thermodynamic coefficients relative to the ensemble of variables $\mathcal{E}_q$ knowing those relative to the ensemble of variables $\mathcal{E}_p$. It is then possible to form a new characteristic matrix relative to ensemble $\mathcal{E}_q$.

NOTE.– The general relations established above are complex and a direct calculation method that involves choosing a variable is preferred, which acts as a function in the two initial and final ensembles, to write the differential in two different ways as shown with relations [2.11] and [2.12].

We will demonstrate this method in the following example: we have selected a phase with one component a move from ensemble $\mathcal{E}_0$ relative to the internal energy U to ensemble $\mathcal{E}_1$ relative to Helmholtz energy:

$$\mathcal{E}_0 = (V,S) \text{ and } \mathcal{E}_1 = (V,T)$$

So, let us assume that the coefficients relative to the phase in ensemble $\mathcal{E}_0$ are as follows:

$$\left(\frac{\partial T}{\partial S}\right)_0, \left(-\frac{\partial P}{\partial S}\right)_0, \left(\frac{\partial T}{\partial V}\right)_0 \text{ and } \left(-\frac{\partial P}{\partial V}\right)_0$$

We aim to calculate the coefficients relative to ensemble $\mathcal{E}_1$, i.e.:

$$\left(\frac{\partial S}{\partial T}\right)_1, \left(-\frac{\partial P}{\partial T}\right)_1, \left(\frac{\partial S}{\partial V}\right)_1 \text{ and } \left(-\frac{\partial P}{\partial V}\right)_1$$

First of all, we express the differential of temperature in the first ensemble of variables:

$$\mathrm{d}T = \left(\frac{\partial T}{\partial S}\right)_0 \mathrm{d}S + \left(\frac{\partial T}{\partial V}\right)_0 \mathrm{d}V \tag{2.22}$$

Since pressure is a function in both phases, we can therefore express its differential in two forms:

$$\mathrm{d}P = \left(\frac{\partial P}{\partial S}\right)_0 \mathrm{d}S + \left(\frac{\partial P}{\partial V}\right)_0 \mathrm{d}V = \left(\frac{\partial P}{\partial T}\right)_1 \mathrm{d}T + \left(\frac{\partial P}{\partial V}\right)_1 \mathrm{d}V \tag{2.23}$$

With dT form equation [2.22] in the second form of dP, we obtain:

$$\mathrm{d}P = \left(\frac{\partial P}{\partial T}\right)_1 \left(\frac{\partial T}{\partial S}\right)_0 \mathrm{d}S + \left(\frac{\partial P}{\partial V}\right)_1 \left(\frac{\partial T}{\partial V}\right)_0 \mathrm{d}V + \left(\frac{\partial P}{\partial V}\right)_1 \mathrm{d}V \tag{2.24}$$

Now let us identify the coefficients of dS and dV. Between the first expression of [2.23] and relation [2.24], we obtain:

– by identifying the coefficients of dS:

$$\left(\frac{\partial P}{\partial T}\right)_1 = \left(\frac{\partial P}{\partial S}\right)_0 \frac{1}{\left(\frac{\partial T}{\partial S}\right)_0}$$

– by identifying the coefficients of dV:

$$\left(\frac{\partial P}{\partial V}\right)_1 = \left(\frac{\partial P}{\partial V}\right)_0 - \left(\frac{\partial P}{\partial S}\right)_0 \left(\frac{\partial T}{\partial V}\right)_0 \frac{1}{\left(\frac{\partial T}{\partial S}\right)_0}$$

The symmetry of the characteristic matrix gives:

$$\left(\frac{\partial S}{\partial P}\right)_1 = \left(\frac{\partial P}{\partial T}\right)_1$$

And we therefore obtain:

$$\left(\frac{\partial S}{\partial T}\right)_1 = \frac{1}{\left(\frac{\partial T}{\partial S}\right)_0}$$

And the four new thermodynamic coefficients are obtained.

2.1.6. *Relations between thermodynamic coefficients and secondary derivatives of the characteristic function*

By deriving the expressions such as in relation [1.8], we find that one thermodynamic coefficient is associated with the secondary derivative of the characteristic function, associated with the same variables, with regard to the definition variable and conjugate variable of the definition function of this thermodynamic coefficient:

$$\frac{\partial^2 \Gamma}{\partial Z_i \partial Z_j} = \varepsilon_{i,i} \frac{\partial \Psi_i}{\partial Z_j} = \varepsilon_{i,i} K_{i,j} \qquad [2.25]$$

2.1.7. *Examples of thermodynamic coefficients: calorimetric coefficients*

Let us consider a reversible elementary transformation, the heat exchanged is expressed as a function of the variables chosen for the problem:

$$\mathrm{d}Q = T\,\mathrm{d}S = \chi_1\,\mathrm{d}x_1 + \chi_2\,\mathrm{d}x_2 + \ldots + \chi_i\,\mathrm{d}x_i + \ldots + \chi_n\,\mathrm{d}x_n$$

The terms χ_i are called the *calorimetric coefficients.*

Therefore the calorimetric coefficients appear, as a factor of T, as thermodynamic coefficients whose definition function is the entropy and the definition variable the corresponding variable x_i.

So for any x variable we obtain:

$$K_{S,x} = \frac{\partial S}{\partial x}$$

And the corresponding calorimetric coefficient is:

$$\chi_x = T\frac{\partial S}{\partial x} = TK_{S,x} \qquad [2.26]$$

For example, in T, $-P$ variables we obtain:

$$\mathrm{d}Q = C_p\,\mathrm{d}T + h\,\mathrm{d}P = T\,\mathrm{d}S$$

And therefore:

$$C_P = T\left(\frac{\partial S}{\partial T}\right)_P \quad \text{and} \quad h = T\left(\frac{\partial S}{\partial P}\right)_T = -T\left(\frac{\partial V}{\partial T}\right)_P \qquad [2.27]$$

2.2. Partial molar variables and thermodynamic coefficients

The following theorem can easily be proven:

THEOREM 5.– The partial molar variables are elements of the characteristic matrix associated with the intensive variables.

We will attempt to prove this theorem for a phase with two components and four variables, its extension to a phase with more variable couples is direct. Let us choose the ensemble of variables $E = (T, -P, n_1, n_2)$. The corresponding characteristic matrix is written:

$$|\mathcal{M}(\mathcal{E})| = \begin{Vmatrix} -1 & 0 & 0 & 0 \\ 0 & -1 & 0 & 0 \\ 0 & 0 & 1 & 0 \\ 0 & 0 & 0 & 1 \end{Vmatrix} \begin{Vmatrix} -\left(\frac{\partial V}{\partial P}\right) & -\left(\frac{\partial S}{\partial P}\right) & -\left(\frac{\partial \mu_1}{\partial P}\right) & -\left(\frac{\partial \mu_2}{\partial P}\right) \\ \left(\frac{\partial V}{\partial T}\right) & \left(\frac{\partial S}{\partial T}\right) & \left(\frac{\partial \mu_1}{\partial T}\right) & \left(\frac{\partial \mu_2}{\partial T}\right) \\ \left(\frac{\partial V}{\partial n_1}\right) & \left(\frac{\partial S}{\partial n_1}\right) & \left(\frac{\partial \mu_1}{\partial n_1}\right) & \left(\frac{\partial \mu_2}{\partial n_1}\right) \\ \left(\frac{\partial V}{\partial n_2}\right) & \left(\frac{\partial S}{\partial n_2}\right) & \left(\frac{\partial \mu_1}{\partial n_2}\right) & \left(\frac{\partial \mu_2}{\partial n_2}\right) \end{Vmatrix} = \begin{Vmatrix} \left(\frac{\partial V}{\partial P}\right) & \left(\frac{\partial S}{\partial P}\right) & \left(\frac{\partial \mu_1}{\partial P}\right) & \left(\frac{\partial \mu_2}{\partial P}\right) \\ -\left(\frac{\partial V}{\partial T}\right) & -\left(\frac{\partial S}{\partial T}\right) & -\left(\frac{\partial \mu_1}{\partial T}\right) & -\left(\frac{\partial \mu_2}{\partial T}\right) \\ \left(\frac{\partial V}{\partial n_1}\right) & \left(\frac{\partial S}{\partial n_1}\right) & \left(\frac{\partial \mu_1}{\partial n_1}\right) & \left(\frac{\partial \mu_2}{\partial n_1}\right) \\ \left(\frac{\partial V}{\partial n_2}\right) & \left(\frac{\partial S}{\partial n_2}\right) & \left(\frac{\partial \mu_1}{\partial n_2}\right) & \left(\frac{\partial \mu_2}{\partial n_2}\right) \end{Vmatrix} \quad [2.28]$$

We obtain expression [2.28] in which the following derivatives arise:

$-\left(\frac{\partial V}{\partial n_1}\right)_{P,T,n_1n_2}$ which is the partial molar volume of component 1, i.e. $\overline{V_1}$;

$-\left(\frac{\partial V}{\partial n_2}\right)_{P,T,n_1n_2}$ which is the partial molar volume of component 2, i.e. $\overline{V_2}$;

$-\left(\frac{\partial S}{\partial n_1}\right)_{P,T,n_1n_2}$ which is the partial molar entropy of component 1, i.e. $\overline{S_1}$;

$-\left(\frac{\partial S}{\partial n_2}\right)_{P,T,n_1n_2}$ which is the partial molar entropy of component 2, i.e. $\overline{S_2}$.

2.3. Common variables and thermodynamic coefficients

We have seen that to define a thermodynamic phase, one must know a certain number of thermodynamic coefficients. There is no table or ensemble of data that directly provide us with the thermodynamic coefficients. These must be deduced from other data. We will now focus on some data from tables which allow the desires thermodynamic coefficients to be obtained.

2.3.1. *State equation*

The first ensemble of information is found, using the state equations. We know that such an equation links temperature, volume, pressure and amount of matter. If we require molar variables, the state equation for one mole takes the general form:

$$F(P,V,T)=0$$

It is clear that knowing this equation provides access to a certain number of thermodynamic coefficients, such as:

$$\left(\frac{\partial P}{\partial T}\right)_V, \left(\frac{\partial P}{\partial V}\right)_T, \left(\frac{\partial V}{\partial T}\right)_P, \left(\frac{\partial V}{\partial P}\right)_T.$$

Thermodynamics does not impose a limit and some of these may be zero (but their signs are still valid for a stable phase as we will see in section 2.5).

However, if we now consider the third principle of thermodynamics, entropy is zero at any pressure at absolute zero, that is to say, at a temperature of 0 K, the derivative of entropy with regard to pressure (at a constant temperature) is zero, and hence at this temperature, the derivative of volume with regard to temperature, which is the symmetrical thermodynamic coefficient in the characteristic matrix $\mathfrak{M}(-P,T)$, is also zero.

NOTE.– Note that the term "state equation" for an equation in the form $F(P,V,T)=0$ is improper for such an equation does not allow a state to be defined, even in the two most simple cases of ensembles (P,T and S,V). In fact, this function will only provide, at a temperature other than absolute zero, two terms of the characteristic matrix, where at least three are required to characterize a whole matrix and therefore the phase.

If we assess the previous state functions linking pressure, volume and temperature, we find that they link the couple of conjugate variables (V and P) to another variable belonging to another couple: temperature. Other state functions have been formed in the same form. The most famous is the magnetic state equation of paramagnetic substances, or the Curie law, which links the couple magnetic moment, $\mathfrak{M}$, and magnetic field, $\mathcal{H}$, at a given temperature for the so-called perfect paramagnetic substances and is written as follows, with C being a constant:

$$\frac{\mathcal{M}}{\mathcal{H}} = \frac{C}{T} \qquad [2.29]$$

The Eotvos law links surface area A, surface energy σ and temperature T, with two constants: critical temperature T_c and constant k:

$$\sigma A = k(T - T_e) \qquad [2.30]$$

NOTE.– The Eotvos law can also take the following form:

$$\sigma\left(\frac{M}{\rho}\right)^{2/3} = k(T - T_e)$$

where M represents the molar mass and ρ the volumic mass of the liquid.

A law of the same type also exists for the stretching of rubber between the force applied F to a cylinder of initial length l_0 and stretching it to l at temperature T through two constants B and the cubic expansion coefficient α (section 2.3.2.1):

$$F = BT\left\{\frac{l}{l_0} - \left[1 + \alpha(T - T_0)\right]\left(\frac{l_0}{l}\right)^2\right\} \qquad [2.31]$$

2.3.2. *Expansion coefficients*

Several common variables, grouped under the term *thermoelastic coefficients*, are defined from the thermodynamic coefficients.

2.3.2.1. *Cubic expansion coefficient (or coefficient of relative increase in volume)*

At a constant pressure, it is defined for by:

$$\beta = \frac{1}{V}\left(\frac{\partial V}{\partial T}\right)_P \qquad [2.32]$$

For this coefficient at the occupied volume V_0 at a temperature of 0°C:

$$\beta_0 = \frac{1}{V_0}\left(\frac{\partial V}{\partial T}\right)_P \quad [2.33]$$

2.3.2.2. *Coefficient of pressure increase at a constant volume*

It is defined by relation:

$$\delta = \frac{1}{P}\left(\frac{\partial P}{\partial T}\right)_V \quad [2.34]$$

Which is sometimes defined, at pressure P_0 at a temperature of 0°C, by:

$$\delta_0 = \frac{1}{P_0}\left(\frac{\partial P}{\partial T}\right)_V \quad [2.35]$$

2.3.2.3. *Isothermal compressibility coefficient at constant temperature T*

It is defined by relation:

$$\chi_T = -\frac{1}{V}\left(\frac{\partial V}{\partial P}\right)_T \quad [2.36]$$

The minus sign has been introduced since, as we know (see section 1.1.1) the true intensive conjugate variable at a volume is the opposite of pressure: $-P$.

NOTE.– If for the considered phase there is a state equation of type $f(P, V, T) = 0$, we obtain relation:

$$\beta = \delta \chi_T P \quad [2.37]$$

In fact, we can write:

$$\delta\chi_T P = -\frac{1}{P}\frac{\frac{\partial f}{\partial P}}{\frac{\partial f}{\partial T}} \cdot \frac{1}{V}\frac{\frac{\partial f}{\partial V}}{\frac{\partial f}{\partial P}} P = -\frac{1}{V}\frac{\frac{\partial f}{\partial V}}{\frac{\partial f}{\partial T}} = \frac{1}{V}\frac{\partial V}{\partial T} = \beta \quad [2.38]$$

2.3.2.4. *Linear expansion coefficient*

It is defined by:

$$\alpha = \frac{1}{l}\left(\frac{\partial l}{\partial T}\right) \qquad [2.39]$$

Sometimes this coefficient at length l_0 at a temperature of 0°C is:

$$\alpha_0 = \frac{1}{l_0}\left(\frac{\partial l}{\partial T}\right) \qquad [2.40]$$

NOTE.– For an isotrope phase, this coefficient is the same in all three directions of space, if g is a shape factor, the volume is always written:

$$V = gl^3$$

i.e.:

$$\left(\frac{\partial V}{\partial T}\right) = 3gl^2\frac{\partial l}{\partial T} = 3\frac{V}{l}\frac{\partial l}{\partial T}$$

and therefore:

$$\beta = 3\alpha \qquad [2.41]$$

2.3.3. *Molar heat capacities*

The original definition of molar heat capacities was defined as the quantity of heat needed to increase the temperature of one mole of substance by 1°C (without any change in phase). In fact, as this heating can be carried out in different ways, reversible or non-reversible, usually the quantity of heat used, which is not a state function, depends on the heating method; it is preferred to define the heat capacities from the entropy function which is a state function according to the second principle.

We define the molar heat capacity at constant volume and the molar heat capacity at constant pressure by the following expressions, respectively:

$$C_V = \frac{1}{T}\left(\frac{\partial S}{\partial T}\right)_V \qquad [2.42]$$

and

$$C_P = \frac{1}{T}\left(\frac{\partial S}{\partial T}\right)_P \qquad [2.43]$$

Thus, the molar heat capacities are linked to two thermodynamic coefficients with temperature as the definition variable and entropy as the definition function.

Specific heat (c_V or c_P) is sometimes used, which represents the same concept, but for one gram of matter instead of one mole, i.e.:

$$C_V = Mc_V \qquad [2.44]$$

or

$$C_P = Mc_P \qquad [2.45]$$

In the case of phase with several components there is, generally, no additivity of heat capacities of different compounds because there is no additivity of entropy.

Using the variable change method (see section 2.1.5) we see that between the heat capacities at constant volume and pressure, we obtain the following relation:

$$C_P - C_V = T\left(\frac{\partial P}{\partial T}\right)_V \left(\frac{\partial V}{\partial T}\right)_P \qquad [2.46]$$

This is a simple change in the ensemble of variables, moving from $(T,-P)$ to ensemble (T,V).

2.3.4. *Young's Modulus*

In the field of elastic deformation, a bar with cross-section s and initial length l_0 and current length l, subject to a traction force F, undergoes stretching and Young's modulus is the variable:

$$\mathcal{E} = \frac{l_0 s}{\dfrac{1}{\left(\dfrac{\partial l}{\partial F}\right)}} \qquad [2.47]$$

In general, this modulus does not need to be defined at constant pressure as this variable has no influence on this phenomenon.

2.3.5. *Electric permittivity*

The charge q of a capacitor is associated with the difference in voltage ϕ through its capacity c according to the equality:

$$q = c\varphi \text{ therefore } c = \frac{\partial q}{\partial \varphi} \qquad [2.48]$$

Now the capacity is proportional to a shape factor g through the electrical permittivity of its dielectric ε:

$$c = \varepsilon g \qquad [2.49]$$

Therefore:

$$\frac{\partial q}{\partial \varphi} = \varepsilon g \qquad [2.50]$$

Since the permittivity and shape of the region or acceleration field are known, the thermodynamic coefficient $\dfrac{\partial q}{\partial \phi}$ is known.

2.3.6. *Volumic and area densities of electric charge*

The volumic ρ and area σ_e densities of electric charges are also defined by the thermodynamic coefficients according to:

$$\frac{\partial q}{\partial V} = \rho \qquad [2.51]$$

and

$$\frac{\partial q}{\partial \mathrm{A}} = \sigma_e \qquad [2.52]$$

2.4. Thermodynamic charts: justification of different types

We have seen that to describe the thermodynamic properties of a phase either a characteristic function or the terms of a characteristic matrix must be known, or a certain number of carefully chosen thermodynamic coefficients. The functions usable for real phases do not always take mathematically simple forms that allow easy manipulation as one often resorts to graphical representations, called thermodynamic charts. There are a large number of possibilities; however, the two main types of chart in use are discussed in the following.

2.4.1. *Representation of a variable as a function of its conjugate*

The first type covers a point that we have already mentioned, namely, in all cases, one must know the variation in each term of the main diagonal of the matrix of thermodynamic coefficients as a function of its definition variable. These terms are, in absolute value, the derivative of a variable with regards to its conjugate in the form:

$$K_{i,i}(\mathcal{E}_p) = \left(\frac{\partial Y_i}{\partial X_i}\right)_{\mathcal{E}_p} \quad \text{or} \quad K_{k,k}(\mathcal{E}_p) = \left(\frac{\partial X_k}{\partial Y_k}\right)_{\mathcal{E}_p}$$

hence the idea of choosing as a system the axes X_i and Y_i or simple monotone functions of X_i and Y_i. On these charts we will trace the curves along which the other variables of ensemble $\mathcal{E}_p$ will be kept constant.

Some examples from this family of charts in thermodynamics with the two couples (T,S) and (P,V) include Andrew's charts ($P = f(V)$), Clapeyron charts ($PV = f(P)$) and entropic charts ($T = f(S)$), as well as deformation charts

($F = f(l)$), magnetization curves ($\mathcal{M} = f(\mathcal{H})$) – charts giving the chemical potential as a function of composition.

For example, we are going to consider entropic charts or (T,S) charts.

Firstly let us consider (Figure 2.1(a)), at constant pressure, we obtain:

$$\mathrm{d}S = \frac{C_P}{T}\mathrm{d}T$$

If C_p varies little with temperature, the integration gives the equation of the isobar at P:

$$S = C_P \ln T + C_1$$

C_1 is independent of temperature and therefore, since the heat capacity at constant pressure varies little with temperature, the isobar is a logarithmic curve and two isotherms are practically parallel.

It is clear that the heat capacity at constant pressure is given by the sub-tangent AB.

Now let us consider the isochoric curves (curves at constant volume). Their equation is given by:

$$\mathrm{d}S = \frac{C_V}{T}\mathrm{d}T$$

Since the heat capacity at constant volume is practically independent of temperature, the integration gives:

$$S = C_V \ln T + C_2$$

C_2 is independent of temperature and therefore, since the heat capacity at constant volume varies little with temperature, the isobar is a logarithmic curve and two isotherms are practically parallel.

It is clear that the heat capacity at constant volume is given by the sub-tangent CD.

Now, if we compare the isobars and isochoric curves, we generally obtain:

$$C_P > C_V$$

We find that the isochoric curves are steeper than the isobars.

Note that since the heat capacities of stable phases are positive (see section 2.5.5), isobaric and isochoric curves are always increasing for stable phases.

Now we are going to show how to obtain different thermodynamic coefficients from the chart. We aim to calculate the different coefficients at point M (characterized, for example, by its temperature and pressure).

Calculating the coefficients $(\partial T / \partial S)_P$, $(\partial T / \partial S)_V$, $(\partial S / \partial T)_P$ and $(\partial S / \partial T)_V$ is clear from the tangents. For example, $(\partial T / \partial S)_P$ is determined simply by considering the tangent at the isobaric curve passing through M (Figure 2.1(a)). We immediately obtain:

$$\left(\frac{\partial T}{\partial S}\right)_P = \frac{\text{MB}}{\text{AB}}$$

The three other coefficients can be easily obtained either on the isobaric curve or on the isochoric curve passing through M.

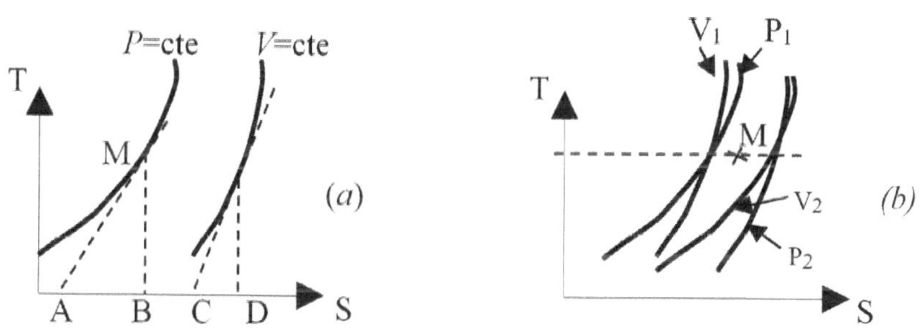

Figure 2.1. *Entropic chart*

With regard to these other coefficients, an interpolation method is used, which we will describe by calculating the coefficient $(\partial V / \partial P)_T$. Close to point M (Figure 2.1b) we see a first couple, an isobar at P_1 and an isochoric curve at V_1, which intersect at the temperature of point M. As close as possible, on the other side of point M, we see a second couple, an isobaric curve at P_2 and an isochoric curve at V_2, which also intersect at the temperature of point M and we write:

$$\left(\frac{\partial V}{\partial P}\right)_T \approx \left(\frac{V_2 - V_1}{P_2 - P_1}\right)_Y$$

To calculate the coefficient $(\partial V / \partial T)_P$, we choose the isobaric curve passing through M and look for two intersections between an isochoric curve and a horizontal line intersecting the chosen isobaric curve either side of M, and we instantly calculate:

$$\left(\frac{\partial V}{\partial T}\right)_P \approx \left(\frac{V_2 - V_1}{T_2 - T_1}\right)_P$$

The method is generalized for all other possible coefficients.

2.4.2. *Representation of a characteristic function as a function of one of its natural variables*

The second type of graphical representation uses as a system of axes a characteristic function and one of its canonical variables; we trace a network of curves, where each one is characterized by the value of the second canonical variable.

Included within this method of representation are Mollier charts, giving the function of enthalpy as a function of entropy ($H - S$) on which isothermal area is found, refrigeration charts giving enthalpy as a function of the logarithm of pressure (H – ln$P$) and physico-chemical charts giving Gibbs free energy as a function of the composition ($G - x$). We also find charts that give the ratio G/T as a function of its composition.

We are now going to determine the thermodynamic coefficients on this type of diagram. We will use the Mollier chart, which is composed of an ensemble of isobaric and isothermic curves.

For example, to calculate the coefficient $(\partial V / \partial S)_P$, we write:

$$\left(\frac{\partial V}{\partial S}\right)_P = \left(\frac{\partial\left(\frac{\partial H}{\partial S}\right)_P}{\partial P}\right)_S$$

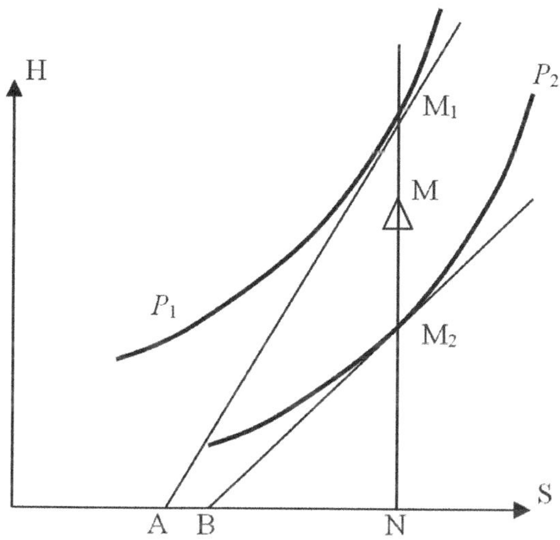

Figure 2.2. *Mollier chart*

We consider (Figure 2.2) two isobaric curves at P_1 and P_2, either side of point M and the two tangents of these isobars at points M_1 and M_2 on the vertical of M (same value of S); these tangents intersect the axis of entropies in A and B, respectively. The vertical line passing through M intersects the X-axis at N and we obtain:

$$\left(\frac{\partial H}{\partial S}\right)_{P_1} = \frac{\text{AN}}{\text{M}_1\text{N}} \quad \text{and} \quad \left(\frac{\partial H}{\partial S}\right)_{P_2} = \frac{\text{BN}}{\text{M}_2\text{N}}$$

$$\text{and} \left(\frac{\partial V}{\partial S}\right)_P \approx \left(\frac{\left(\frac{\partial H}{\partial S}\right)_{P_1} - \left(\frac{\partial H}{\partial S}\right)_{P_2}}{P_2 - P_1} \right)_S = \frac{\frac{\text{BN}}{\text{M}_2\text{N}} - \frac{\text{AN}}{\text{M}_1\text{N}}}{P_2 - P}$$

We will proceed in the same manner for the other coefficients, either from the isobaric or isothermal curves, or both as shown above.

2.5. Stability of phases

For a phase to be stable, its thermodynamic coefficients must satisfy certain conditions.

2.5.1. *Case of ensemble $\mathcal{E}_0$ of extensive variables*

$\mathcal{E}_0$ is defined as the ensemble of extensive variables in couples of variables appropriate for the problem. One mole of substance can be found in two states:

– a first state, where the mole is fully included in a phase (α);

– a second state, where the mole is distributed between phase (α) and phase (β) adjacent to the first. Between these two phases, the molar variables only differ by an infinitesimal quantity.

In the second state, the phase (α) has $1 - \delta\xi$ moles and phase (β) $\delta\xi$ moles. ξ is the fractional extent or transfer. The extensive molar values are:

$$\frac{X_k}{n} = \overline{x_k}$$

such that in the first state:

$$(1 - \delta\alpha)\delta^{(\alpha)}\left(\overline{x_k}\right) + \delta^{(\alpha)}\left(\overline{x_k}\right)\delta^{(\beta)}\left(\overline{x_k}\right) = 0$$

In the second state:

$$X_k = (1 - \delta\xi)\left[\overline{x_k} + \delta^{(\alpha)}\left(\overline{x_k}\right)\right] + \delta\xi\left[\left(\overline{x_k}\right) + \delta^{(\beta)}\left(\overline{x_k}\right)\right]$$

Now assuming X_k is the same in both states, we obtain:

$$(1-\delta\alpha)\delta^{(\alpha)}\left(\overline{x_k}\right)+\delta^{(\alpha)}\left(\overline{x_k}\right)\delta^{(\beta)}\left(\overline{x_k}\right)=0 \qquad [2.53]$$

We therefore obtain p relations (p is the number of couples of variables involved) relations of type [2.53] with $1 \leq k \leq p$.

The internal energy in the first state is:

$$U_1 = u^{(\alpha)} \qquad [2.54]$$

In the second state, the internal energy is:

$$(1-\delta\alpha)\delta^{(\alpha)}(x_k)+u^{(\beta)}\delta\xi \qquad [2.55]$$

Therefore:

$$\delta U = -u^{(\alpha)}\delta\xi + u^{(\beta)}\delta\xi \qquad [2.56]$$

With U being a potential function and the transformation of phase (α) into phase (β) not being spontaneous for phase (α) to be stable, we must have:

$$\frac{\mathrm{d}U}{\mathrm{d}\xi} > 0$$

We obtain δU by calculating $u^{(\alpha)}$ and $u^{(\beta)}$.

By developing $u^{(\alpha)}$ into a Mac Laurin series, we obtain:

$$u^{(\alpha)} = u + \sum_{k=1}^{p}\frac{\partial u}{\partial \overline{x_k}}\delta^{(\alpha)}\left(\overline{x_k}\right)+\frac{1}{2}\sum_{j=1}^{p}\sum_{k=1}^{p}\frac{\partial^2 u}{\partial \overline{x_k}\partial \overline{x_j}}\delta^{(\alpha)}\left(\overline{x_k}\right)\delta^{(\alpha)}\left(\overline{x_j}\right) \qquad [2.57]$$

And the same for $u^{(\beta)}$:

$$u^{(\beta)} = u + \sum_{k=1}^{p} \frac{\partial u}{\partial x_k} \delta^{(\beta)}\left(\overline{x_k}\right) + \frac{1}{2} \sum_{j=1}^{p} \sum_{k=1}^{p} \frac{\partial^2 u}{\partial x_k \partial x_j} \delta^{(\beta)}\left(\overline{x_k}\right) \delta^{(\beta)}\left(\overline{x_j}\right) \quad [2.58]$$

Now according to expression [2.53], we obtain:

$$\delta^{(\alpha)}\overline{x_k} = -\frac{\delta\xi}{1-\delta\xi} \delta^{(\beta)}\left(\overline{x_k}\right) \quad [2.59]$$

By including the equalities [2.57], [2.58] and [2.59] in expression [2.55], it becomes:

$$\delta U = \frac{\delta\xi}{2(1-\delta\xi)} \sum_{j=1}^{p} \sum_{k=1}^{p} \frac{\partial^2 u}{\partial x_k \partial x_j} \delta^{(\beta)}\left(\overline{x_k}\right) \delta^{(\beta)}\left(\overline{x_j}\right)$$

By dividing by $\delta\xi$ and with $\delta\xi$ tending towards zero, we obtain:

$$\frac{\mathrm{d}U}{\mathrm{d}\xi} = \lim_{\delta\xi \to 0} \frac{1}{2} \sum_{j=1}^{p} \sum_{k=1}^{p} \frac{\partial^2 u}{\partial x_k \partial x_j} \delta^{(\beta)}\left(\overline{x_k}\right) \delta^{(\beta)}\left(\overline{x_j}\right)$$

The stability condition of the phase therefore becomes:

$$\sum_{j=1}^{p} \sum_{k=1}^{p} \frac{\partial^2 u}{\partial x_k \partial x_j} \delta^{(\beta)}\left(\overline{x_k}\right) \delta^{(\beta)}\left(\overline{x_j}\right) > 0$$

The thermodynamic coefficients are the secondary derivatives of the function U typical for the ensemble of variables E_0 and therefore:

$$\frac{\partial^2 u}{\partial x_k \partial x_j} = K_{j,k}(\mathcal{E}_0) = K_{k,j}(\mathcal{E}_0)$$

Which gives the stability condition of the phase (α), whether it be $\partial \overline{x_j}$ or $\partial \overline{x_k}$:

$$\sum_{j=1}^{p}\sum_{k=1}^{p} K_{j,k}(\mathcal{E}_0)\,\delta^{(\beta)}\left(\overline{x_k}\right)\delta^{(\beta)}\left(\overline{x_j}\right) > 0 \qquad [2.60]$$

Let us consider the specific couple (j, k), its contribution to the previous sum is:

$$f_{j,k} = \frac{1}{2}K_{j,j}(\mathcal{E}_0)\left[\delta^{(\beta)}\left(\overline{x_j}\right)\right]^2 + \frac{1}{2}K_{k,k}(\mathcal{E}_0)\left[\delta^{(\beta)}{''}\left(\overline{x_k}\right)\right]^2 + K_{j,k}\delta^{(\beta)}\left(\overline{x_k}\right)\delta^{(\beta)}\left(\overline{x_j}\right) > 0 \qquad [2.61]$$

For the sum to be positive, the contribution of each couple must be positive and therefore:

$$f_{j,k} > 0 \qquad [2.62]$$

With $f_{j,k}$ being a quadratic formula, it will be positive if the three following conditions are fulfilled:

$$K_{j,j}(\mathcal{E}_0) > 0 \qquad [2.63]$$

$$K_{k,k}(\mathcal{E}_0) > 0 \qquad [2.64]$$

$$K_{j,j}(\mathcal{E}_0)K_{k,k}(\mathcal{E}_0) > K_{k,j}(\mathcal{E}_0)K_{j,k}(\mathcal{E}_0) \qquad [2.65]$$

So it appears that matrix $\mathfrak{M}(\mathcal{E}_0)$, which is symmetrical, is defined as positive.

2.5.2. *Coefficients associated with ensemble $\mathcal{E}_n$*

The ensemble of variables E_n is defined by the ensemble of intensive variables selected as variables of the problem, i.e.:

$$\mathcal{E}_n = \{Y_1, Y_2, \ldots Y_n\}$$

Matrix $\left\| K_{i,j}(\mathcal{E}_n) \right\|$ is the inverse of matrix $\left\| K_{j,i}(\mathcal{E}_0) \right\|$, i.e.:

$$\mathrm{d}Y_i = K_{i,j}(\mathcal{E}_0)\mathrm{d}X_i \qquad [2.66]$$

and

$$\mathrm{d}X_j = K_{j,i}(\mathcal{E}_n)\mathrm{d}Y_i \qquad [2.67]$$

Therefore:

$$K_{j,i}(\mathcal{E}_n) = \left[K_{i,j}(\mathcal{E}_0) \right]^{-1} \qquad [2.68]$$

The inverse of a positively defined matrix is itself positive, i.e.:

– for *i*, we obtain:

$$K_{i,i}(\mathcal{E}_n) > 0 \qquad [2.69]$$

– for any *i,j* couple, we obtain:

$$K_{i,i}(\mathcal{E}_n)K_{j,j}(\mathcal{E}_n) > K_{i,j}(\mathcal{E}_n)K_{j,i}(\mathcal{E}_n) \qquad [2.70]$$

Note that matrix $\left\| K_{i,j}(\mathcal{E}_n) \right\|$ is also symmetrical since for *i* we obtain:

$$\varepsilon_{i,i}(\mathcal{E}_n) = -1$$

2.5.3. *Case of other ensembles of variables*

Through the method developed in section 2.1.5 we can move from ensemble $\mathcal{E}_n$ to ensemble $\mathcal{E}_{n-1}$ defined by:

$$\mathcal{E}_{n-1} = \{X_1, Y_2, \ldots Y_n\}$$

We therefore establish the relations:

$$K_{i,1}(\mathcal{E}_n) = K_{i,1}(\mathcal{E}_{n-1})K_{1,1}(\mathcal{E}_n) \qquad [2.71]$$

$$K_{i,j}(\mathcal{E}_n) = K_{i,1}(\mathcal{E}_{n-1})K_{1,j}(\mathcal{E}_n) + K_{i,j}(\mathcal{E}_{n-1}) \quad [2.72]$$

with:

$$K_{j,1}(\mathcal{E}_{n-1}) = -K_{1,j}(\mathcal{E}_{n-1}) \quad [2.73]$$

and:

$$K_{i,j}(\mathcal{E}_{n-1}) = -K_{j,i}(\mathcal{E}_{n-1}) \quad \text{for } j \neq 1 \text{ and } j \neq i \quad [2.74]$$

$$K_{1,1}(\mathcal{E}_{n-1}) = \frac{1}{K_{1,1}(\mathcal{E}_n)} \quad [2.75]$$

From the first relation we obtain, taking into account equality [2.70]:

$$K_{1,1}(\mathcal{E}_{n-1}) > 0 \quad [2.76]$$

From expressions [2.71] and [2.73] we deduce:

$$K_{i,i}(\mathcal{E}_{n-1}) = \frac{K_{i,i}(\mathcal{E}_n)K_{1,1}(\mathcal{E}_n) - K_{i,1}(\mathcal{E}_n)K_{1,i}(\mathcal{E}_n)}{K_{1,1}(\mathcal{E}_n)} \quad [2.77]$$

By applying expression [2.69] with $j = 1$, we obtain:

$$K_{i,1}(\mathcal{E}_{n-1}) > 0 \quad [2.78]$$

According to relation [2.73]:

$$K_{1,i}(\mathcal{E}_{n-1})K_{i,1}(\mathcal{E}_{n-1}) = -\left[K_{1,i}(\mathcal{E}_{n-1})\right]^2 \quad [2.79]$$

Now according to relation [2.78] we obtain:

$$K_{1,i}(\mathcal{E}_{n-1})K_{j,j}(\mathcal{E}_{n-1}) > 0 \quad [2.80]$$

Therefore:

$$K_{i,i}(\mathcal{E}_{n-1})K_{1,1}(\mathcal{E}_{n-1}) > K_{i,1}(\mathcal{E}_{n-1})K_{1,i}(\mathcal{E}_{n-1}) \quad [2.81]$$

Now if we apply relation [2.72] to inequality [2.70] for i and j different to 1, we obtain:

$$K_{j,j1}(\mathcal{E}_{n-1})K_{i,i}(\mathcal{E}_{n-1}) > \left[K_{1,j}(\mathcal{E}_{n-1})\right]^2 + AK_{1,1}(\mathcal{E}_n) \quad [2.82]$$

A is a quadratic form defined by:

$$A = K_{j,j}(\mathcal{E}_{n-1})\left[K_{1,i}(\mathcal{E}_{n-1})\right]^2 + K_{i,i}(\mathcal{E}_{n-1})\left[K_{1,j}(\mathcal{E}_{n-1})\right]^2$$
$$-2K_{i,j}(\mathcal{E}_{n-1})K_{1,i}(\mathcal{E}_{n-1})K_{1,j}(\mathcal{E}_{n-1})$$

Condition [2.82] will be satisfied if:

$$K_{j,j}(\mathcal{E}_{n-1})K_{i,i}(\mathcal{E}_{n-1}) > \left[K_{i,j}(\mathcal{E}_{n-1})\right]^2 \quad [2.83]$$

with $A > 0$.

Now, for A to be positive, condition [2.83] must also be satisfied. Condition [2.83] is the only condition that needs to be satisfied for phase (α) to be stable.

2.5.4. *Conclusion: stability conditions of a phase in terms of thermodynamic coefficients*

Step by step, we can move from ensemble $\mathcal{E}_{n-1}$ to $\mathcal{E}_{n-2}$, etc., hence the following theorem:

THEOREM 6.– The stability of a phase requires the following conditions on the thermodynamic coefficients (whatever the ensemble of variables $\mathcal{E}_p$ chosen):

$$\begin{cases} K_{i,i}(\mathcal{E}_p) > 0 & [2.84] \\ K_{i,i}(\mathcal{E}_p)K_{j,j}(\mathcal{E}_p) > K_{i,j}(\mathcal{E}_p)K_{j,i1}(\mathcal{E}_p) & [2.85] \end{cases}$$

2.5.5. *Example – applying stability conditions*

We are now going to take the typical simple case of an ensemble of two variables $\mathcal{E}(-P,T)$, the matrix of thermodynamic coefficients is:

$$\mathcal{M}(\mathcal{E}) = \begin{Vmatrix} \left(\dfrac{\partial S}{\partial T}\right)_P & \left(\dfrac{\partial V}{\partial T}\right)_P \\ \left(-\dfrac{\partial S}{\partial P}\right)_T & \left(-\dfrac{\partial V}{\partial P}\right)_T \end{Vmatrix}$$

Applying the relation [2.84] immediately gives $(\partial S/\partial T)_P > 0$ and therefore $C_P > 0$ (with the help of another ensemble $\mathcal{E}(V,T)$ we now also prove that the heat capacity at constant volume is also positive), i.e. it is impossible to cool down a substance by supplying heat.

For the second term of the main diagonal, we obtain:

$$\left(\frac{\partial V}{\partial P}\right)_T < 0$$

It is impossible to compress a substance by decreasing the pressure.

Applying the inequality [2.85], we obtain:

$$-\left(\frac{\partial V}{\partial P}\right)_T\left(\frac{\partial S}{\partial T}\right)_P > -\left(\frac{\partial V}{\partial T}\right)_P\left(\frac{\partial S}{\partial P}\right)_T$$

The symmetry of the matrix gives:

$$\left(\frac{\partial S}{\partial P}\right)_T = -\left(\frac{\partial V}{\partial T}\right)_P$$

and hence the third condition:

$$\left(\frac{\partial V}{\partial T}\right)_P^2 < -\left(\frac{\partial S}{\partial T}\right)_P \left(\frac{\partial V}{\partial P}\right)_T$$

This condition is deduced by introducing the heat capacity and thermoelastic coefficients by:

$$\left(\frac{\partial V}{\partial T}\right)_P^2 < -\frac{C_P}{T}\left(\frac{\partial V}{\partial P}\right)_T \qquad [2.86]$$

For example, for a perfect gas the condition is represented by:

$$\mathrm{R} < C_P$$

We can also represent the inequality [2.86] in terms of thermomechanic coefficients, i.e.:

$$V^2\alpha^2 < \frac{C_P}{T}\chi$$

2.6. Consistency of thermodynamic data

When solving a thermodynamic problem, it is also important that the thermodynamic data, which can come from multiple sources, are consistent with one another. Similarly, when building a model, the laws obtained must also fulfill certain rules of consistency (one example of this is provided by the Debye-Hückel equation for modeling ionic solutions). According to what we have seen in this chapter, the conditions this data must satisfy are:

– conditions of symmetry of the characteristic matrix (and not of the thermodynamic coefficients) (section 2.1.4);

– the thermodynamic coefficients of the main diagonal of the matrix of thermodynamic coefficients must be positive (section 2.5.4);

– relations of inequality between non-diagonal thermodynamic coefficients (section 2.5.4);

– particular relations between certain thermodynamic coefficients, i.e. the Gibbs–Duhem equation must be applied to thermodynamic coefficients that are partial molar variables (section 1.2);

– relations between thermodynamic coefficients with regard to ensembles of different variables (section 2.1.5).

Under these conditions the following are added:

– extensive variables must be homogenous of 1 degree with regard to quantities of matter, and the intensive variables must be homogenous of zero degree with regard to the quantities of matter (see Appendix 3, section A3.1);

– the partial molar variables must be intensive.

Verifying all these properties allows us to determine whether the thermodynamic data collected for a phase are consistent with one another.

2.7. Conclusion on the macroscopic modeling of phases

Therefore from the "basic" experimental data such as a state equation or thermoelastic coefficients, heat capacities etc., it is possible to obtain thermodynamic coefficients. A double integration of equation [2.25] will give a characteristic function, i.e. knowing the complete thermodynamics of the phase studied.

3

Multi-Compound Phases – Solutions

Systems with several components contained within a single phase are called:

– solutions, in the case of condensed liquid or solid phases;

– mixtures, in the case of gases that are miscible in all proportions.

NOTE.– we do not use the term "mixture" for condensed solutions since a mixture of liquids or solids may contain one or several phases, and liquids and solids are not always miscible in all proportions. This means that a system may occur whose composition is not homogeneous; it is therefore composed of several phases. (Note that a phase is a region of space in which all intensive variables are continuous functions of space variables.)

3.1. Variables attached to solutions

The involvement of several components in a phase consequently introduces new couples of intensive and extensive variables (one couple per component): chemical potentials and amounts of matter (see section 2.2.4), and then the involvement of new variables: compositions, partial molar variables (see section 2.3), chemical potentials (see section 2.6) and mixing variables.

3.1.1. *Characterizing a solution*

Knowing a solution means being able to express the thermodynamic coefficients or a characteristic function in at least one ensemble of variables. To study the solutions we often use the ensemble of variables consisting of:

– physical intensive p variables (temperature, pressure, other interesting intensive variables for studying and using the phase, such as electric field, surface tension, etc.);

– amounts of matter of each component expressed as number of moles.

The system will be completely characterized if the function's generalized Gibbs energy, G_G (relation [1.9]) is known, which is written, for a solution with N components whose generalized chemical potentials are $[\mu_i]$, in the differential form:

$$dG_G = -\sum_{j=1}^{p} X_j \, dY_j + \sum_{i=1}^{N} [\mu_i] \, dn_i \qquad [3.1]$$

To determine this function, the generalized chemical potentials must be known as these potentials are the generalized partial Gibbs energies, i.e.:

$$G_G = \sum_{i=1}^{N} [\mu_i] n_i \qquad [3.2]$$

With regard to chemical systems (with the only intensive variables: temperature and pressure) the characteristic function is the Gibbs energy G which obeys the corresponding relations involving the chemical potential:

$$dG = -S\,dT + P\,dV + \sum_{i=1}^{N} \mu_i \, dn_i \qquad [3.3]$$

$$G = \sum_{i=1}^{N} \mu_i n_i \qquad [3.4]$$

In this chapter, we will not distinguish between a generalized system and a chemical system, since the relations are the same in both cases.

3.1.2. *Composition of a solution*

The composition of a solution gives the quantities and proportions of each component. In addition to expressing the quantities in numbers of moles, other variables are used to express the composition of a solution. We will now look at the main ones.

3.1.2.1. *Molar fractions*

The molar fraction x_i of a component i in a phase, which contains n_i moles of compound i in a total of N components, is the ratio of the amount of compound i to the total amount of matter in the phase, i.e.:

$$x_i = \frac{n_i}{\sum_{i-1}^{N} n_i} \text{ with } 0 \leq x_i \leq 1 \text{ and } \sum_{i=1}^{N} x_i = 1 \qquad [3.5]$$

3.1.2.2. *Mass fractions*

The mass fraction m_i of a component i in a phase is the ratio of mass of this component to the total mass of the phase, i.e.:

$$w_i = \frac{m_i}{\sum_{i=1}^{N} m_i} = \frac{m_i}{m} \qquad [3.6]$$

Note that we can vary the molar fraction or mass fraction of a component without adding or subtracting from this component, but simply by modifying the amount of another component. All molar fractions and mass fractions of a solution vary when a certain amount of a component is added or subtracted.

3.1.2.3. *Concentrations or molarities*

The concentration (or molarity) C_i of a component i in a phase is the quotient of the amount of this component per volume of the phase, i.e.:

$$C_i = \frac{n_i}{V} \qquad [3.7]$$

Concentration is most commonly expressed in moles/liter. For solutions in which a component is in large proportions with regard to the others (often called the solvent), the volume of the phase s is often confused with the introduced volume of the most plentiful component. The concentration of this solvent is often considered to be constant if a small amount of another component is added.

3.1.2.4. *Molalities*

The molality M_i of a component *i* with a low content in a phase (often called the solute) is the quotient of the quantity of this component per mass of main component *0* (often called the solvent), i.e.:

$$\mathsf{M}_i = \frac{n_i}{m_0} \qquad [3.8]$$

Clearly it is easy to pass from one variable to another through the intermediary of molar masses and/or volumic masses of different components in the solution.

3.1.3. *Peculiar variables and mixing variables*

An extensive variable, such as the volume of a solution, is not the simple addition of volumes of substances introduced when a solution is being constituted. Usually, a term is added only due to the mixing of substances.

3.1.3.1. *Definitions*

Let us consider a partial molar variable $\overline{J_i}$. In a uniform solution, this variable is a function of the variables of the problem, i.e. some intensive and extensive variables belong to other conjugate couples of *J*. If we have one mole of a solution overall, by having the molar fractions x_i as the composition variables, we can write:

$$\overline{J_i} = \overline{J_i}(Y_1, Y_2 ... Y_p, x_1, x_2, ..., x_N) \qquad [3.9]$$

It is always possible to break down this function into the sum of two functions such that one only depends on the intensive variables called the *peculiar function* $\overline{J_i}^{pec}$, and the other is the function of all variables, particularly composition, called the *mixing function* $\overline{J_i}^{mix}$:

$$\overline{J_i} = \overline{J_i}^{pec}(Y_1, Y_2, ...Y_p) + \overline{J_i}^{mix}(Y_1, Y_2, ...Y_p, x_1 x_2 ... x_N) \qquad [3.10]$$

This decomposition is not unequivocal, there are an infinite number of couples that satisfy the relation [3.10]. To define the decomposition chosen, additional information is needed.

Often the molar value of the pure body i in the same aggregation state as the solution is chosen as the peculiar value, which means that if the solution is a gaseous mixture, i must be gaseous, since under the conditions chosen, the substance i does not exist in the gaseous state. The same goes for a liquid solution; pure i must be chosen as a liquid. For a solid solution, i must be chosen as a solid and in the same crystalline phase as the solution, even if this phase does not exist. The reference in the same condensation state as the solution is called the *conservation of aggregation state*. The decomposition [3.10] is therefore written:

$$\overline{J_i} = j_i^0(Y_1, Y_2, ...Y_p) + \overline{J_i}^{mix}(Y_1, Y_2, ...Y_p, x_1 x_2 ... x_N) \qquad [3.11]$$

j_i^0 designates the molar value of the variable for pure substance i in the same aggregation state as the solution.

3.1.3.2. *Molar variables of mixing*

The operation [3.11] can also be applied to any linear combination of partial molar variables and particularly to the combination:

$$J_i = \sum_{i=1}^{N} n_i \overline{J_i} \qquad [3.12]$$

We call J_m^{mix} the molar variable of mixing of a solution defined by:

$$J_m^{mix} = \sum_{i=1}^{N} x_i \left(J_i - j_i^0\right) \qquad [3.13]$$

This definition of the variable of mixing is applied to any extensive variable. Let us consider the following examples.

3.1.3.3. *Molar Gibbs energy of mixing*

According to relation [3.13], the molar Gibbs energy of mixing is defined by:

$$G_m^{mix} = \sum_{i=1}^{N} x_i\left(\overline{G_i} - g_i^0\right) = \sum_{i=1}^{N} x_i\left(\mu_i - g_i^0\right) \qquad [3.14]$$

As the partial molar Gibbs energy is confused with the chemical potential, the corresponding partial molar variables for each substance in the solution will be in the form:

$$\overline{G_i}^{mix} = \mu_i - g_i^0 \qquad [3.15]$$

3.1.3.4. *Other molar variables of mixing*

Knowing the molar Gibbs energy of mixing, it is easy to determine all other variables. In particular, for the molar enthalpy of mixing, we obtain:

$$H_m^{mix} = -\frac{\partial\left(\dfrac{G^{mix}}{T}\right)}{\partial T} = \sum_{i=1}^{N} x_i\left(\overline{H_i} - h_i^0\right) \qquad [3.16]$$

And for the corresponding partial molar values:

$$\overline{H_i}^{mix} = \overline{H}_i - h_i^0 = -\frac{\partial\left(\dfrac{\overline{G_i}^{mix}}{T}\right)}{\partial T} \qquad [3.17]$$

Similarly, for the molar entropy of mixing, we obtain:

$$S_m^{mix} = \sum_{i=1}^{N} x_i\left(\overline{S_i} - s_i^0\right) = -\frac{\partial G^{mix}}{\partial T} = \frac{H^{mix} - G^{mix}}{T} \qquad [3.18]$$

And for the partial molar entropy:

$$\overline{S_i}^{mix} = \overline{S}_i - s_i^0 = \frac{\overline{H_i}^{mix} - \overline{G_i}^{mix}}{T} \qquad [3.19]$$

Variables of mixing are defined from the molar variables of pure components, taken as reference states.

3.2. Recap of ideal solutions

The concept of an ideal solution forms the basis of the macroscopic modeling of all solutions. We are now going to describe some macroscopic properties of an ideal solution.

3.2.1. *Thermodynamic definition*

We define an ideal solution as a solution in which the chemical potential of each component obeys the following relation:

$$\mu_i = g_i^0 + \mathrm{R}T\, ln\, x_i \qquad [3.20]$$

From this definition we can deduce a certain number of characteristic properties of an ideal solution.

3.2.2. *Molar Gibbs energy of mixing of an ideal solution*

By using relation [3.14] we obtain:

$$G_m^{mix} = \sum_{i=1}^{N} x_i \left(\mu_i - g_i^0\right) = \mathrm{R}T \sum_{i=1}^{N} x_i\, ln x_i \qquad [3.21]$$

3.2.3. *Molar enthalpy of mixing of the ideal solution*

By using relation [3.16] we obtain:

$$H_m^{mix} = \sum_{i=1}^{N} x_i \left(\overline{H_i} - h_i^0\right) = -\frac{\partial\left(\dfrac{G_m^{mix}}{T}\right)}{\partial T} = 0 \qquad [3.22]$$

Therefore, the molar Gibbs enthalpy of mixing of an ideal solution is zero at all temperatures and all pressures. This means that the components of an ideal solution mix without the need for heat.

The same result is obtained for internal energy. The internal energy of mixing of an ideal solution is zero at all temperatures and all pressures.

3.2.4. *Molar entropy of mixing of the ideal solution*

By using relation [3.18] we obtain:

$$S_m^{mix} = \frac{H_m^{mix} - G_m^{mix}}{T} = -R\sum_{i=1}^{N} x_i\, lnx_i \qquad [3.23]$$

The entropy of mixing of an ideal solution is not zero; however, it does not depend on temperature.

3.2.5. *Molar volume of mixing*

We are going to calculate the volume of mixing from the following expression:

$$V_m^{mix} = \frac{\partial G_m^{mix}}{\partial P} = \sum_{i=1}^{N} x_i\left(\overline{V}_i - v_i^0\right) \qquad [3.24]$$

Taking into account expression [3.21] we obtain:

$$V_m^{mix} = 0 \qquad [3.25]$$

Therefore the compounds of the ideal solution mix without any change in volume. This means that the volume of one mole of the ideal solution is the sum, weighted per molar fraction, of molar volumes of pure compounds. This result is valid at all temperatures and all pressures.

3.2.6. *Molar heat capacity of ideal solution: Kopp's law*

The molar heat capacity can be calculated, by considering relation [3.22], using:

$$C_P^{mix} = \frac{\partial H_m^{mix}}{\partial T} = 0 \qquad [3.26]$$

Therefore the heat capacity of mixing of the ideal solution at constant pressure is zero, therefore heat capacity of the solution at constant pressure is written:

$$C_P = \sum_{i=1}^{N} c_{P(i)}^0 \qquad [3.27]$$

This relation constitutes Kopp's law.

NOTE.– Of the properties of the ideal solution that we have just mentioned, only one, the Gibbs energy of mixing, accepts the reciprocal. In fact, Gibbs energy is the characteristic function of the system in P,T variables. However, for the other functions that we have examined, the fact that one of them, for a given solution, has the same property as that of an ideal solution, does not mean that the studied solution is perfect. This is the case, in particular, for so-called *regular* solutions which have the same entropy of mixing as a perfect solution, without being one.

3.3. Characterization imperfection of a real solution

The properties of real solutions considerably differ from those of an ideal solution. It was found to be necessary to obtain a mean, if possible, instead of relying solely on this approximation and to characterize real solutions, to better describe their properties.

Two methods have emerged, both of which characterize a real solution from its difference to an ideal solution:

– the first method, the Lewis method, introduces the activity coefficients in the expression of the chemical potential in a way that preserves the form of the law of mass action of chemical equilibrium. This method is mainly used for calculating equilibria between phases of the same substance and those of simple chemical equilibria;

– the second method involves variables of excess and in particular the excess Gibbs energy. This is the difference between the Gibbs energy of a solution and that of an ideal solution under the same conditions. This second method is mainly used when building microscopic models of solutions gives the excess Gibbs energy. It is also widely used for calculating complex equilibria which is done, through computer calculations, by minimizing the Gibbs energy function.

Clearly these two forms are ultimately equivalent via a simple expression, which we will establish later (expression [3.83]), which links the activity coefficients of different components of the solution and their excess partial molar Gibbs energies.

3.3.1. *Lewis activity coefficients*

Lewis refers to expression [3.20] of the chemical potential by trying to conserve its form. The following are introduced: the coefficient of activity γ, a function of temperature, pressure, etc. and composition of the solution, by writing the chemical composition of the component i of the solution in the form:

$$\mu_i = \mu_i^0 + RT\,ln\,\gamma_i x_i \qquad [3.28]$$

3.3.1.1. *Coefficients of activity and reference states*

The product of the coefficient of activity and the molar fraction is called the *activity* of component i in the solution. The activity in a real solution will play the same thermodynamic role as the molar fraction in an ideal solution:

$$a_i = \gamma_i x_i \qquad [3.29]$$

The expression [3.28] introduces two functions: μ_i^0, called the *reference chemical potential* and coefficient of activity γ_i. In fact, this introduction requires another piece of data. The reference state definition is chosen by means of conventions.

In practice, several conventions are introduced; three of which are frequently used, which we denote convention (I), (II) or (III). Each one has a corresponding reference chemical potential, respectively, $\mu_i^{0(I)}$, $\mu_i^{0(II)}$ or $\mu_i^{0(III)}$. Each convention will also have a corresponding coefficient of activity $\gamma_i^{(I)}$, $\gamma_i^{(II)}$ or $\gamma_i^{(III)}$ and an activity denoted $a_i^{(I)}$, $a_i^{(II)}$ or $a_i^{(III)}$. Each time an expression is valid, whatever the reference, we omit the distinctive sign (I), (II) or (III).

The value of the chemical potential of a component in the solution does not depend on the reference chosen, because, as the chemical potential is the partial molar Gibbs energy, the Gibbs energy of the solution would depend on it, which would be absurd.

Convention (I) selects as a reference the components in a pure state in the same aggregation state as the solution (i.e. the pure compound reference). This is what we selected in section 3.1.3.2 for variables of mixing. Under these conditions, the chemical potential of the reference state is the molar Gibbs energy of the pure compound i and the chemical potential is therefore written:

$$\mu_i = g_i^0 + RT \ln \gamma_i^{(I)} x_i \qquad [3.30]$$

This reference is used especially when the components of a solution play the same role and have similar molar fractions. For example, this convention is chosen when data covers a wide composition spectrum, ultimately passing from one pure substance to another.

Convention (II) (called the infinitely dilute solution reference) distinguishes amongst the components of a solution, those in the largest proportion(s), which are called the solvent(s) and those in the lowest proportions, which are called the solutes. The reference state is different for both categories of components:

– for a solvent, its pure state is chosen (in the same aggregation state as the solution) as a reference and therefore the chemical potential will obey relation [3.30]; convention (II) is then confused with convention (I) for a solvent;

– for a solute, the reference state is an imaginary solution in which all the solutes are infinitely dilute. The chemical potential of the reference is therefore that of the solute in this imaginary solution denoted μ_S^{∞} for a solute s. The coefficient of activity of the solute s in this imaginary solution is equal to 1. For solutes, the activity tends towards 1 as the molar fraction tends towards zero; the chemical potential is therefore written:

$$\mu_s = \mu_s^{\infty} + RT \ln \gamma_s^{(II)} x_s \qquad [3.31]$$

The coefficients of activity with reference solute (I) (pure compound) and with reference solute (II), (infinitely dilute solution) are linked since the chemical potential of the solute does not depend on the chosen convention. We can therefore write:

$$\mu_s = g_s^0 + RT\,ln\,\gamma_s^{(I)} x_s = \mu_s^\infty + RT\,ln\,\gamma_s^{(II)} x_s \qquad [3.32]$$

From which we can deduce:

$$ln\frac{\gamma_s^{(I)}}{\gamma_s^{(II)}} = \frac{\mu_s^\infty - g_s^0}{RT} = ln\,K_{iH} \qquad [3.33]$$

The constant K_{iH} linking the coefficients of activity expressed in the two conventions: pure compound and infinitely dilute solution, for a solute, does not depend on the solution composition but does depend on the values of the intensive variables (pressure, temperature etc.) by means of chemical potentials of the reference states. The variable K_{iH} is called *Henry's law constant* (see section 3.4).

We have seen that for a solvent, both conventions are identical.

NOTE.– In a solution with more than two components, nothing prevents us from considering several solvents and several solutes.

The convention (III) distinguishes between the solvents and solutes:

– for the solvent, the reference convention (III) is identical to the reference (I) pure compound, and therefore the chemical potential of the solvent is always given by relation [3.30];

– for a solute, we write that the coefficient of activity is equal to 1 for a solution containing 1 mole per liter of each solute. The chemical potential of the reference state is that of the solute in this solution containing 1 mole/l of each solute. The chemical potential of the solute is therefore written:

$$\mu_s = \mu_s^{0(III)} + RT\,ln\,\gamma_s^{(III)} C_s \qquad [3.34]$$

NOTE.– It is not impossible, that in some cases of low solubility, the reference solution with one mole per liter of each solute is imaginary since it is impossible to create. This does not detract from the fact that the convention requires only that the solvent and solutes of the reference solution be in the same aggregation state as the solution, even if it is an imaginary state.

Convention (III), molar reference solution, is mainly used for ionic aqueous solutions, even though in the previous case, the reference chosen is often the solution whose molarity tends towards zero, which does not change the chemical potential of the reference state $\mu_s^{0(III)}$.

There is clearly a relation between the coefficients of activity expressed in this convention (III) and those from other conventions. For the solvent, this relation is clear since the three conventions are identical; therefore, all the coefficients of activity are equal regardless of the convention chosen.

For solutes, we have already seen the relation between the coefficients of activity in conventions (I) and (II) (relation [3.33]). Let us examine the relation between the coefficients of activity expressed using the diluted reference solution (II) and those from the molar solution convention (III).

By defining the molar fractions (relation [3.5]), by assigning 0 to the solvent and s to the solutes, we can write:

$$\frac{x_s}{x_0} = \frac{n_s}{n_0} \quad [3.35]$$

So, the volume of the solution is, according to the partial molar volumes of the components, given by:

$$V = n_0 \overline{V_0} + \sum_s n_s \overline{V_s} \quad [3.36]$$

And therefore by using relation [3.7] it becomes:

$$C_s = \frac{n_s}{n_0 \overline{V_0} + \sum_s n_s \overline{V_s}} = \frac{x_s}{x_0 \overline{V_0} + \sum_s x_s \overline{V_s}} \quad [3.37]$$

But if the solution is sufficiently dilute and if v_0^0 designates the molar volume of the pure solvent, we can write:

$$\sum_s x_s \overline{V_s} << x_0 \overline{V_0} \cong \overline{V_0} \cong v_0^0 \quad [3.38]$$

The molarity can therefore be approximately written as:

$$C_s \cong \frac{x_s}{v_0^0} \tag{3.39}$$

The chemical potential does not depend on the reference state chosen and by considering relation [3.39], we obtain:

$$\mu_s = \mu_s^{0(III)} + \mathrm{R}T \ln \gamma_s^{(III)} C_s \cong \mu_s^{\infty} + \mathrm{R}T \ln \gamma_s^{(II)} C_s + \mathrm{R}T \ln v_0^0 \tag{3.40}$$

Hence, by identifying:

$$\mu_s^{0(III)} \cong \mu_s^{\infty} + \mathrm{R}T \ln v_0^0 \quad \text{and} \quad \gamma_s^{(III)} \cong \gamma_s^{(II)} \tag{3.41}$$

So in a sufficiently diluted solution, the coefficients of activity, in the conventions infinitely dilute solution reference (II) and molar solution reference (III), are identical for both the solvent as well as the solutes.

3.3.1.2. *Relation between the coefficients of activity of the components of a solution*

For any given convention, we know that according to the Gibbs–Duhem equation, with constant external intensive variables (temperature, pressure, etc.), the chemical potentials obey the Gibbs–Duhem equation [1.24].

By dividing by the total number of moles n, we obtain:

$$\sum_{i=1}^{N} x_i \, d\mu_i = 0 \tag{3.42}$$

By differentiating expression [3.28] with constant external intensive variables, we obtain:

$$d\mu_i = \mathrm{R}T \, d \ln \gamma_i x_i \tag{3.43}$$

So, applied to inequality [3.42] we obtain:

$$\sum_{i=1}^{N} x_i d \ln \gamma_i x_i = \sum_{i=1}^{N} x_i d \ln \gamma_i + \sum_{i=1}^{N} x_i d \ln \gamma_i = 0 \tag{3.44}$$

Now it is clear that according to relation [3.5] we obtain:

$$\sum_{i=1}^{N} x_i d\, ln\, x_i = 0 \qquad [3.45]$$

From which we deduce:

$$\sum_{i=1}^{N} x_i d\, ln\, \gamma_i = 0 \qquad [3.46]$$

Thus, if we know for each solution composition, the coefficients of activity of all the components in a convention, except one of them, relation [3.46] allows the unknown coefficient of activity to be calculated in the same convention.

3.3.1.3. *Influence of temperature on the coefficients of activity*

Let us consider chemical systems.

Expression [3.28] of the chemical potential of component i of the solution can be written as follows:

$$\frac{\mu_i}{T} = \frac{\mu_i^0}{T} + \mathrm{R}\, ln\, \gamma_i + \mathrm{R}\, ln\, x_i \qquad [3.47]$$

According to the second Helmholtz relation [1.31]:

$$\frac{\partial\left(\frac{\mu_i}{T}\right)}{\partial T} = -\frac{\overline{H_i}}{T^2} \qquad [3.48]$$

Similarly, this relation can be applied to the reference state of the solution, i.e.:

$$\frac{\partial\left(\frac{\mu_i^0}{T}\right)}{\partial T} - \frac{\overline{H_i^0}}{T^2} \qquad [3.49]$$

$\overline{H_i^0}$ is the partial molar Gibbs energy of component i in the reference state, at temperature T. By inserting expressions [3.48] and [3.49] into the equality [3.47] after derivation, we obtain:

$$\frac{\partial \ln \gamma_i}{\partial T} = \frac{\overline{H_i^0} - \overline{H_i}}{RT^2} \quad [3.50]$$

This relation [3.50] shows that the coefficient of activity varies with temperature, at constant pressure and composition.

The numerator of the right-hand side of expression [3.50] generates a difference, which represents the variation in Gibbs energy when one mole of *i* is transported, at constant pressure and temperature, from the real solution studied into the solution in the reference state. We must imagine that a large amount of solution is transported without the compositions of either solution being modified. If we look at convention (I), it is clear that this difference is the inverse of the Gibbs energy of mixing.

3.3.1.4. *Influence of other intensive variables on the coefficients of activity*

If we choose an intensive variable *Y*, conjugate of the extensive variable *X*, we know that the variations in generalized chemical potential of a component *i* with this variable *Y* is in the form:

$$\frac{\partial [\mu_i]}{\partial Y} = -\overline{X_i} \quad [3.51]$$

$\overline{X_i}$ is the partial molar variable corresponding to variable *X*.

Applying the same relation to the reference solution, we obtain:

$$\frac{\partial \left[\mu_i^0\right]}{\partial Y} = -\overline{X_i^0} \quad [3.52]$$

By including relations [3.51] and [3.52] in relation [3.28], after derivation we obtain:

$$\frac{\partial \ln \gamma_i}{\partial Y} = \frac{\overline{X_i^0} - \overline{X_i}}{RT} \quad [3.53]$$

Such a relation can be applied, for example, to the influence in pressure in a chemical system in the form:

$$\frac{\partial \ln \gamma_i}{\partial P} = -\frac{\overline{V_i^0} - \overline{V_i}}{RT} \qquad [3.54]$$

$\overline{V_i}$ and $\overline{V_i^0}$ are the partial molar volumes of component i in the real solution and in the reference solution, respectively.

3.3.1.5. *Another expression of the chemical potential in a condensed solution*

If we consider a condensed solution, liquid or solid, and we define, by analogy with the coefficient of compressibility of a phase, the coefficient at constant temperature is:

$$\overline{\chi_i} = -\frac{1}{\overline{V_i}} \frac{\partial \overline{V_i}}{\partial P} \qquad [3.55]$$

Let us assume that this independent variable, pressure, is small enough that the usual pressures are always:

$$\overline{\chi_i} P << 1 \qquad [3.56]$$

By integration, the expression [3.55] becomes:

$$\overline{V_i} = \overline{V_i^0}\left(1 - \overline{\chi_i} P\right) \qquad [3.57]$$

$\overline{V_i^0}$ is the partial molar volume of component i extrapolated at zero pressure.

By applying relation [3.52] to the variable pressure (-*P*), it becomes:

$$\frac{\partial \mu_i^0}{\partial P} = \overline{V_i^0} \qquad [3.58]$$

And by integration at pressures between 0 and *P*, we obtain:

$$\mu_i^0(T,P) - \mu_i^0(T,0) = \overline{V_i} P\left(1 - \frac{\overline{\chi_i} P}{2}\right) \qquad [3.59]$$

i.e.

$$\mu_i^0(T,0)=\eta_i(T) \qquad [3.60]$$

With $\eta_i(T)$ being a function of temperature, relation [3.59] is written:

$$\mu_i(T,P)=\eta_i(T)+\overline{V_i^0}P\left(1-\frac{\overline{\chi_i}P}{2}\right)+RT\ln\gamma_i x_i \qquad [3.61]$$

This relation separates the variables pressure and temperature in the expression of the chemical potential of the component *i* in the reference state. This separation is used for condensed solutions, which are often considered to be incompressible except at very high pressure, where relation [3.61] is applied. For gases, this separation of variables is expressed in a different manner.

3.3.1.6. *Influence of composition on the coefficients of activity*

The thermodynamic coefficients, whose definition function is a chemical potential and the variable, the amount of a component, are of two types:

– diagonal terms, for which the chemical potential and amount of matter are relative to the same component: $\partial\mu_i\ /\partial n_i$. We know that these terms are positive if the phase is stable (see section 2.5.4);

– symmetrical terms, of the form $\partial\mu_i\ /\partial n_k$ for which the chemical potential is relative to the component *i* and the amount of matter is relative to another component *k*. These terms are equal two by two (matrix symmetry), i.e.:

$$\frac{\partial\mu_i}{\partial n_k}=\frac{\partial\mu_k}{\partial n_i} \qquad [3.62]$$

First of all, let us examine the symmetrical terms; by derivation we can write:

$$\frac{\partial\mu_i}{\partial n_k}=\frac{\partial\mu_i}{\partial x_i}\frac{\partial x_i}{\partial n_k}+\frac{\partial\mu_i}{\partial x_k}\frac{\partial x_k}{\partial n_k}+\sum_{j\neq i,k}\frac{\partial\mu_i}{\partial x_j}\frac{\partial\mu_j}{\partial n_i} \qquad [3.63]$$

Now, by deriving the relation [3.28], we obtain:

$$\frac{\partial \mu_i}{\partial x_i} = \mathrm{R}T\left(\frac{\partial \ln \gamma_i}{\partial x_i} + \frac{1}{x_i}\right) \quad [3.64]$$

and $$\frac{\partial \mu_i}{\partial x_k} = \mathrm{R}T\left(\frac{\partial \ln \gamma_i}{\partial x_k} - \frac{1}{x_i}\right) \quad [3.65]$$

So $x_i = \frac{n_i}{n}$ therefore $\frac{\partial x_i}{\partial n_i} = \frac{n - n_i}{n^2}$ and $\frac{\partial x_i}{\partial n_j} = -\frac{n_j}{n^2}$ [3.66]

By referring to expression [3.65], we obtain:

$$\frac{\partial \mu_i}{\partial x_k} = \frac{\mathrm{R}T}{n}\left(\frac{\partial \ln \gamma_i}{\partial x_k} - 1 \sum_{j=1}^{N} x_j \frac{\partial \ln \gamma_i}{\partial x_j}\right) \quad [3.67]$$

If we apply the symmetry relation [3.62], we can write:

$$\frac{\partial \ln \gamma_i}{\partial x_k} - \sum_{j=1}^{N} x_j \frac{\partial \ln \gamma_i}{\partial x_j} = \frac{\partial \ln \gamma_k}{\partial x_i} - \sum_{j=1}^{N} x_j \frac{\partial \ln \gamma_k}{\partial x_j} \quad [3.68]$$

For this identity to always be satisfied, regardless of the type of solution or composition, it is necessary that:

$$\sum_{j=1}^{N} x_j \frac{\partial \ln \gamma_i}{\partial x_j} = \sum_{j=1}^{N} x_j \frac{\partial \ln \gamma_k}{\partial x_j} = 0 \quad [3.69]$$

This is only possible if, for each couple *i, k*, we have:

$$\frac{\partial \ln \gamma_i}{\partial x_k} = \frac{\partial \ln \gamma_k}{\partial x_i} \quad [3.70]$$

And by referring to equality [3.67], we obtain:

$$\frac{\partial \mu_i}{\partial n_k} = \frac{\mathrm{R}T}{n}\left(\frac{\partial \ln \gamma_i}{\partial x_k} - 1\right) \quad [3.71]$$

Now, let us examine the diagonal terms:

$$\frac{\partial \mu_i}{\partial n_i} = \frac{\partial \mu_i}{\partial x_i}\frac{\partial x_i}{\partial n_i} + \sum_{j \neq i} \frac{\partial \mu_i}{\partial x_j}\frac{\partial x_j}{\partial n_i} \quad [3.72]$$

So, by proceeding as above and by taking relation [3.68] into account, we obtain:

$$\frac{\partial \mu_i}{\partial n_i} = \frac{\mathrm{R}T}{n}\left(\frac{\partial \, ln \gamma_i}{\partial x_i} + \frac{1}{x_i} - 1\right) \quad [3.73]$$

We observe that, according to expressions [3.71] and [3.73], the thermodynamic coefficients associated with the chemical potentials and amounts of matter, are known if we know the variations in coefficients of activity with the molar fractions.

3.3.1.7. *Ideal solutions and coefficients of activity*

We defined an ideal solution from relation [3.20], i.e. stating that the coefficient of activity in convention (I) (pure reference compound) was equal to 1 at all temperatures, all pressures and all compositions. We will show that this causes all coefficients of activity, regardless of the convention chosen, to be equal to 1 for an ideal solution.

Indeed, if the component i is a solvent, since its coefficient of activity does not depend on the convention, we obtain:

$$\gamma_0^{(I)} = \gamma_0^{(II)} = \gamma_0^{(III)} = 1 \quad [3.74]$$

If component i is a solute, designated by s, according to relation [3.46], we obtain:

$$x \mathrm{d} \, ln \gamma_0^{(II)} + \sum x_s \mathrm{d} \, ln \gamma_s^{(II)} = 0 \quad [3.75]$$

For this equality to be satisfied in all cases, we require that:

$$\mathrm{d} \, ln \gamma_s^{(II)} = 0 \text{ i.e. } \ln \gamma_s^{(II)} = \text{constant} \quad [3.76]$$

But if x_s tends towards 0, $\gamma_s^{(II)}$ tends towards 1 and therefore the constant of relation [3.76] equals 1:

$$\gamma_s^{(II)} = 1 \qquad [3.77]$$

In the same way we show that:

$$\gamma_s^{(III)} = 1 \qquad [3.78]$$

Thus, for an ideal solution, all coefficients of activity of all components are equal to 1 whatever convention used.

NOTE.– The reciprocal of this property is not true. In a solution, the property $\gamma_s^{(II)} = 1$ does not necessarily give $\gamma_s^{(I)} = 1$ and therefore the solution is not necessarily ideal.

3.3.2. *Characterizing the imperfection of a real solution by the excess Gibbs energy*

A second method, which is used most often, to define a solution, is to define its difference from the ideal solution by the excess Gibbs energy.

3.3.2.1. *Definition of excess variables*

Let us consider an extensive variable J of a solution; we call J^* the value this variable would have under the given conditions of temperature and pressure if this solution was ideal. We call this excess value of J the variable J^{xs} defined by:

$$J^{xs} = J - J^* \qquad [3.79]$$

J^{xs} is characteristic of the difference between our real solution and an ideal solution.

With this variable being extensive, it therefore corresponds to partial molar variables, relative to each component of the solution, defined by:

$$\overline{J^{xs}} = \left(\frac{\partial J^{xs}}{\partial n_i}\right)_{P,T,n_{j\neq i}} \qquad [3.80]$$

We can use this definition for different extensive variables and we will do this for Gibbs energy, entropy, enthalpy and heat capacity.

3.3.2.2. *Excess Gibbs energy*

For a solution, the excess Gibbs energy will be according to the definition:

$$G^{xs} = G - G^* \qquad [3.81]$$

So, the molar Gibbs energy of the solution can be written:

$$G_m = \sum_{i=1}^{N} x_i \mu_i = \sum_{i=1}^{N} \left[x_i \left(g_i^0 + \mathrm{R}T \ln x_i + \mathrm{R}T \ln \gamma_i^{(I)} \right) \right] \qquad [3.82]$$

Similarly, for an ideal solution, this molar Gibbs energy would be:

$$G_m^* = \sum_{i=1}^{N} \left[x_i \left(g_i^0 + \mathrm{R}T \ln x_i \right) \right] \qquad [3.83]$$

From this we deduce the excess molar Gibbs energy of the studied solution:

$$G_m^{xs} = \mathrm{R}T \sum_{i=1}^{N} x_i \ln \gamma_i^{(I)} \qquad [3.84]$$

If the excess Gibbs energy is known, the solution is completely characterized. If we add the Gibbs energy of the ideal solution (relation [3.21]) and those corresponding to the Gibbs energies of each pure compound, we obtain the Gibbs energy of the solution which is the characteristic function with our choice of variables *P*, *T* and composition.

By applying relation [3.80] to the Gibbs energy, it immediately becomes, for the excess partial molar Gibbs energy of component i:

$$\overline{G_i^{xs}} = \mathrm{R}T\left(ln\gamma_i^{(I)} + \sum_{i=1}^{N} x_j \frac{\partial\, ln\gamma_j^{(I)}}{\partial n_i} \right) \qquad [3.85]$$

So, from relation [3.46] we deduce:

$$\sum_{i=1}^{N} x_j \frac{\partial\, ln\gamma_j^{(I)}}{\partial n_i} = 0 \qquad [3.86]$$

Thus defining excess partial molar Gibbs energy of component i:

$$\overline{G_i^{xs}} = \mathrm{R}T\, ln\gamma_i^{(I)} \qquad [3.87]$$

The relation [3.87] links the solutions described by the Lewis method of coefficients of activity and the excess Gibbs energy method. Both methods are therefore equivalent.

3.3.2.3. *Excess entropy*

We know that entropy is linked to Gibbs energy by the relation:

$$S = -\frac{\partial G}{\partial T} \qquad [3.88]$$

By applying this relation to excess variables, for the excess molar entropy, it becomes:

$$S_m^{xs} = \mathrm{R}\sum_{i=1}^{N}\left[x_i \left(-T\frac{\partial \ln\gamma_i^{(I)}}{\partial T} - \ln\gamma_i^{(I)} \right)\right] \qquad [3.89]$$

And for the excess partial molar entropy of component i:

$$\overline{S_i^{xs}} = \mathrm{R}\left(-T\frac{\partial \ln\gamma_i^{(I)}}{\partial T} - \ln\gamma_i^{(I)} \right) \qquad [3.90]$$

This relation can also be obtained by deriving the relation [3.89] with regard to the quantity n_i as well as by applying relation [3.88] to expression [3.87].

3.3.2.4. *Excess enthalpy*

To calculate enthalpy, we will use either the definition of Gibbs entropy, or the first Helmholtz relation [1.29]. We obtain the following expressions: for excess molar enthalpy:

$$H_m^{xs} = -\mathrm{R}\sum_{\mathrm{i=1}}^{\mathrm{N}}\left[x_i T^2 \frac{\partial \ln \gamma_i^{(I)}}{\partial T}\right] \tag{3.91}$$

And for the excess partial molar entropy of component *i*:

$$\overline{H_i^{xs}} = -\mathrm{R}T^2 \frac{\partial \ln \gamma_i^{(I)}}{\partial T} \tag{3.92}$$

3.3.2.5. *Excess heat capacity*

By using the definition of molar heat capacity at constant pressure:

$$C_p = T\left(\frac{\partial S_m}{\partial T}\right) \tag{3.93}$$

By using expression [3.89] we can calculate:

$$C_p^{xs} = \mathrm{R}\sum_{\mathrm{i=1}}^{\mathrm{N}}\left[x_i\left(-T\frac{\partial^2 \ln \gamma_i^{(I)}}{\partial T^2} - 2\frac{\partial \ln \gamma_i^{(I)}}{\partial T}\right)\right] \tag{3.94}$$

Of course, other excess variables, such as for example excess volume, can be calculated from relation [3.84].

3.3.3. *Other ways to measure the imperfection of a solution*

We have seen two equivalent methods to characterize the imperfection of a solution: via the coefficients of activity and the excess Gibbs energy. On rare occasions, we find other methods; we will briefly describe two of them in this section.

3.3.3.1. *The osmotic coefficient method*

This method essentially applies to solutions in which a distinction is made between the solvent and solutes. When taking into account an imperfection, the following relation is written for a solvent [3.30]; in place of the activity coefficient in the logarithm, we can introduce a corrective term according to the logarithm and write:

$$\mu_0 = g_0^0 + g_0 RT \ln x_0 \qquad [3.95]$$

g_0 is called the *osmotic coefficient* of the solvent.

Of course $g_0 \to 1$ if $x_0 \to 1$.

If we consider a dilute solution with only one solute s in the previous solvent, we can use the following approximation:

$$\ln x_0 = \ln(1 - x_s) \cong -x_s \qquad [3.96]$$

By comparing this with relation [3.30] we obtain:

$$\ln \gamma_0^{(II)} x_0 = -g_0 x_s \qquad [3.97]$$

This relation shows the equivalence between the coefficients of activity and osmotic coefficient methods to describe a real solution.

3.3.3.2. *The coefficient method Φ*

If the solution is not dilute, the relation [3.97] is no longer valid, but we can still express the activity in an equivalent way by writing:

$$\ln \gamma_0^{(II)} x_0 = -\Phi_0 x_s \qquad [3.98]$$

Φ_0 is called the *practical osmotic coefficient* relative to the solvent. It is a function of different variables, pressure, temperature and composition. So, we obtain $\Phi_0 \to g_0$ if $x_s \to 0$.

3.4. Activity of a component in any solution: Raoult's and Henry's laws

Let us consider component i in a solution whose molar fraction x varies between 0 and 1 at a given temperature and pressure. In the pure compound reference, the activity of this compound will be given by:

$$a_i^{(I)} = \gamma_i^{(I)} x_i \quad [3.99]$$

If we make x tend towards 1, we know that $\gamma_i^{(I)}$ tends towards 1. According to expression [3.20], a solution, for which the coefficient of activity (convention (I)) equals 1, is an ideal solution. Therefore, when the molar fraction of a component tends towards 1, the behavior of this component tends towards that of an ideal solution; this is what we call *Raoult's law*:

$$\lim_{x_i \to 1} a_i^{(I)} = x_i \quad [3.100]$$

Figure 3.1 shows a curve representing the variation in activity of a component of a solution with composition. Raoult's law for which the activity (convention (I)) is equal to the molar fraction is represented by the bisector at the origin called *Raoult's straight line*, tangent to the curve for $x = 1$.

Now let us examine the behavior of the same activity of i when the molar fraction of i tends towards zero. The solution then behaves with regard to i as an ideal dilute solution. Such a solution is defined by a coefficient in reference (II) which equals 1. Its coefficient of activity in reference (I) will therefore be given by relation [3.33]; the activity of i in convention (I) in an infinitely dilute ideal solution is:

$$\lim_{x_i \to 0} a_i^{(I)} = K_{iH} x_i \quad [3.101]$$

Henry's constant K_{iH} is independent to the composition and therefore has a well-determined value when pressure and temperature are fixed. So, when the molar fraction of a component tends towards zero, its activity tends towards a value proportional to its molar fraction, this is *Henry's law*. This law is represented by *Henry's straight line* in Figure 3.1, a straight line with a slope of K_{iH} .

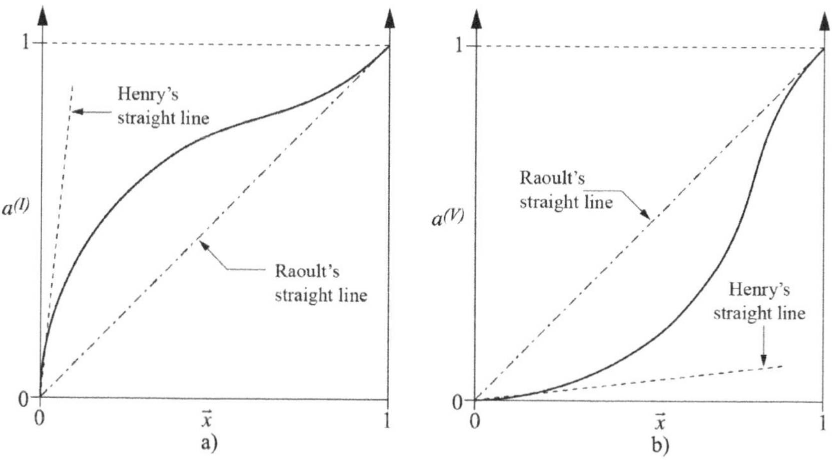

Figure 3.1. *Activity of a constituent in a solution- a) with a positive deviation with regard to Raoult – b) with a negative deviation*

The solution that corresponds to the curve in Figure 3.1(a) is called a *positive deviation* with regard to Raoult's straight line *(*$\gamma^{(l)} > 1$) and a *negative deviation* with regard to Henry's straight line. However, the curve in Figure 3.1(b) corresponds to a solution with a negative deviation with regard to Raoult's straight line *(*$\gamma^{(l)} < 1$) and positive with regard to Henry's straight line. There are also solutions for which the activity curve intersects Raoult's straight line moving from solution with a positive deviation with regard to Raoult's straight line to a solution with a negative deviation with regard to Henry's straight line or vice versa.

A tendency of molecules in components 1 and 2 to attract leads to solutions with a negative deviation with regard to Raoult's law ($\gamma^{(l)} < 1$) whereas the tendency of molecules to repel leads to solutions with a positive deviation with regard to Raoult's law ($\gamma^{(l)} > 1$).

3.5. Ionic solutions

Ionic solutions have several differences compared to molecular solutions whose main source comes from inter-ionic forces greater than those that exist between molecules, particularly over long distances.

3.5.1. *Chemical potential of an ion*

If a solution contains ions next to neutral molecules (of a solvent for example), these ions are components of the system and we must assess their chemical potentials. We know that the chemical potential is confused with the partial molar Gibbs energy and because of this, it is obtained from calculating a partial derivative with regard to the number of ions of a species keeping constant the temperature, pressure and also the amounts of other ion species present in the solution. In the case of charged particles, such as ions, it is impossible to vary the number of ion species by keeping the number of all other ion species constant since the solution will no longer be electrically neutral.

This challenge is not unique to the chemical potential but can also be found for all partial molar variables that call on the same type of partial derivatives. We are going to attempt to circumvent this issue.

For this, let us consider a solution of potassium chloride in water. Macroscopic thermodynamics, which is not concerned with the state of potassium chloride in water, helps us define the following, without ambiguity:

– the chemical potential of the solvent (water) : μ_0 ;

– the chemical potential of the solute, potassium chloride μ_S .

The Gibbs energy of the solution being:

$$G = n_0\mu_0 + n_s\mu_s \quad [3.102]$$

This Gibbs energy is perfectly defined at a given pressure and temperature.

Now let us imagine that our solution is composed of the following species:

– the solvent, water, with the chemical potential μ_0 ;

– Cl^- ions amount n_2 ;

– K^+ ions also amount n_2 .

The partial derivation of the definition of a partial molar variable introduces a virtual term in the chemical potential; this can be written in the following form

$$\varsigma_i = \varsigma_i^0 + \mathrm{R}T \ln \gamma_i C_i + t_i \qquad [3.103]$$

t_i represents an undetermined term, γ_i is the activity of ion i (choosing for example convention III). The Gibbs energy of the solution is therefore written:

$$G = n_1\mu_1 + n_2\left(\varsigma_{K^+}^0 + \varsigma_{Cl^-}^0 + RT\ln \gamma_{K^+} C_{K^+} \gamma_{Cl^-} C_{Cl^-}\right) + n_2\left(t_{Cl^-} + t_{K^+}\right) \qquad [3.104]$$

The value of this Gibbs energy, perfectly defined by expression [3.102], is real. This means that the imaginary terms are eliminated and therefore:

$$n_2\left(t_{Cl^-} + t_{K^+}\right) = 0 \qquad [3.105]$$

It is as if we had retained, as an expression of the chemical potential, the relation:

$$\mu_i = \mu_i^0 + \mathrm{R}T \ln \gamma_i C_i \qquad [3.106]$$

Although this expression may be false it can be used in all linear combinations generating real values such as Gibbs energies.

For ionic solutions, it claims to express chemical potentials in terms of molar concentrations (or molarities), but we have seen that for sufficiently dilute solutions, the molar concentrations are proportional to the molar fractions (see expression [3.39]); we can therefore replace the concentrations with molar fractions provided that the chemical potential of the reference state is appropriately modified. The downside of this substitution is that it can only occur for insufficiently diluted solutions, which makes coefficients of activity variable even for an ideal solution. However, ionic solutions stop behaving perfectly well before the approximation [3.39] becomes no longer tolerable and the coefficients of activity will be variable regardless whether the values are modified or not.

We will assume that the coefficient of activity of an ion should tend towards 1 when its concentration tends towards zero, i.e.:

$$\gamma \to 1 \text{ if } c \to 0 \tag{3.107}$$

3.5.2. *Relation between the activities of ions and the overall activity of solutes*

Let us consider placing an electrolyte with the chemical formula $A_{\nu_+}B_{\nu_-}$ into solution. Its Gibbs energy is given by the relation [3.102] but also by applying relations [3.103] and [3.106], this energy is also written:

$$G = n_0\mu_1 + n_s\left(\nu_+\mu_A + \nu_-\mu_{0_A}\right) \tag{3.108}$$

By identifying the two expressions [3.102] and [3.108] of G, we obtain:

$$\mu_s = \nu_+\mu_A + \nu_-\mu_B \tag{3.109}$$

We can write an equivalent relation for the reference variables:

$$\mu_s^0 = \nu_+\mu_A^0 + \nu_-\mu_B^0 \tag{3.110}$$

Thus deducing:

$$\gamma_s C_s = \left(\gamma_A C_A\right)^{\nu_+}\left(\gamma_B C_B\right)^{\nu_-} \tag{3.111}$$

This equation links the molecular point of view to the ionic point of view.

3.5.3. *Mean concentration and mean ionic activity coefficient*

Methods to measure the activity coefficients cannot deduce individual ionic activity coefficients due to the indetermination that we mentioned above (see section 3.5.1). It is also worth introducing, for an electrolyte $A_{\nu_+}B_{\nu_-}$, the concept of the mean activity coefficient, defined by:

$$\gamma_\pm = \left(\gamma_+^{\nu_+}\gamma_-^{\nu_-}\right)^{1/(\nu_+ + \nu_-)} \tag{3.112}$$

We can also define the mean concentration by an equivalent relation. If C is the molar concentration of the solute, the concentrations of different ions (for completely dissociated strong electrolytes) will be:

$$C_+ = \nu_+ C \text{ and } C_- = \nu_- C \qquad [3.113]$$

And the mean concentration will be:

$$C_\pm = C\left(\nu_+^{\nu_+} \nu_-^{\nu_-}\right)^{1/(\nu_+ + \nu_-))} \qquad [3.114]$$

The mean activity coefficient obeys the same convention [3.107] as the individual activity coefficients.

In particular, for a 1:1 electrolyte such as KCl, we obtain $\nu_+ = \nu_- = 1$ and the previous expressions take the following forms:

$$C_+ = C_- = C = C_\pm^{1/2} \qquad [3.115]$$

$$\gamma_\pm = \left(\gamma_+ \gamma_-\right)^{1/2} \qquad [3.116]$$

In the case non-completely ionized electrolytes, there is the simultaneous presence of molecules and ions, linked by the ionization equilibrium:

$$A_{\nu_+} B_{\nu_-} \Leftrightarrow \frac{\nu_+}{z_A^+} A^{z_A^+} + \frac{\nu_-}{z_B^-} A^{z_B^-}$$

Applying the law of mass action gives:

$$a^{\frac{\nu_+}{z_A^+}} a^{\frac{\nu_-}{z_B^-}} = K \qquad [3.117]$$

Constant K is fixed at a given pressure and temperature.

As activities are only defined by a close proportionality factor, Lewis proposed a convention for non-ionized molecules such that the equilibrium constant K is always equal at 1 at a given temperature and pressure. Lewis showed that it did not matter whether the ionization levels were known since the differences in Gibbs energy had the same value as the molecule whether ionized or not. Therefore the activity can be considered to equally represent both that of non-ionized parts and that of the entire solute. Here, weak electrolytes are treated as above in the case of strong electrolytes.

3.5.4. *Obtaining the activity coefficient of an individual ion*

We have already shown that the experiment only allows us to determine the mean activity coefficients. If we could determine the activity coefficient of a single ion, we could deduce that of all other ions step-by-step. To do this, Mac Innes benefitted from the fact that in potassium chloride, the chloride and the potassium ion have the same charge in absolute value, the same electronic structure and therefore the same size and the same mobility but slightly different masses, and therefore worked out that they should have, in the same solution, the same activity coefficient. From the mean activity coefficient of potassium chloride, we can deduce those of individual ions using the relation:

$$\gamma_{\mathrm{Cl}^-} = \gamma_{\mathrm{K}^+} = \gamma_{\pm}^2 \qquad [3.118]$$

The results show that monovalent ions have roughly the same activity coefficient and that the activity coefficients are influenced by the electrovalence of ions and by the presence of other ions giving rise to what we call ionic strength.

3.5.5. *Ionic strength*

We have just seen that the activity coefficient of ions was influenced by the presence of other ions. The experiment shows that it is not necessarily the concentration that is influenced but a variable introduced by Lewis and Randall: the ionic strength.

This *ionic strength* is defined by the relation:

$$I = \frac{1}{2}\sum_i C_i z_i^2 \qquad [3.119]$$

The sum is extended to all ions presents in the solution.

Note that if the solution only contains a salt $A_{\nu_+}B_{\nu_-}$, the ionic strength is proportional to the concentration, i.e.:

$$I = \frac{1}{2}\left(\nu_+ z_A^2 + \nu_- z_B^2\right)C \qquad [3.120]$$

This introduction allowed Lewis and Randall to determine the values of the activity coefficients of some individual ions, in different solutions. Some are given in Table 3.1. Me^{2+} designates a bivalent cation with a metal origin such as alkaline earth metals, magnesium, copper, cadmium or zinc.

We see that the values of the activity coefficients of the ions essentially depend on the ionic strength of the medium and the electrovalence of the ion.

Nature of the ion	*Ionic strength (moles/l)*		
	0.001	0.01	0.1
H^+	0.98	0.92	0.84
OH^-	0.98	0.92	0.81
Cl^-, Br^-, I^-	0.98	0.92	0.79
Na^+	0.98	0 .92	0.80
K^+, Rb^+, Cs^+	0.98	0.92	0.79
Me^{2+}	0.78	0.60	0.34
SO_4^{2-}	0.77	0.56	0.26
ClO_3^-, BrO_3^-, IO_3^-	0.95	0.85	0.61

Table 3.1. *Values of activity coefficients of individual ions*

3.6. Curves of molar variables as a function of the composition in binary systems of a solution with two components

We are now going to assess the properties of the curve representing a molar variable (volume, entropy, enthalpy, Gibbs energy etc.) of a solution with two components, according to the composition of this solution. As an example, we will take the Gibbs energy.

Let us consider a solution with two components A and B. x is the molar fraction of component B; that of component A is therefore 1-x. We know that the partial molar Gibbs energy is the chemical potential. The molar Gibbs energy of this solution, reduced to one mole of the mixture, is therefore given, according to expression [3.4], by relation:

$$G_m = x\mu_B + (1-x)\mu_A \qquad [3.121]$$

By differentiating this expression and by taking into account the Gibbs–Duhem equation at a fixed pressure and temperature, we obtain:

$$dG_m = \mu_B\, dx + \mu_A\, d(1-x) \qquad [3.122]$$

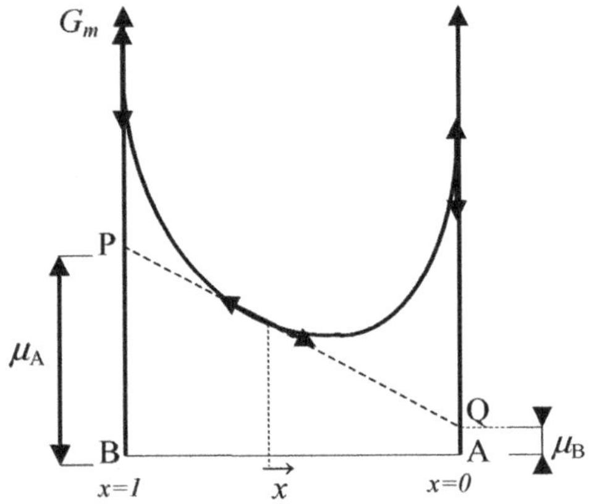

Figure 3.2. *Gibbs energy of a binary solution according to the composition*

Figure 3.2 shows a curve representing the molar Gibbs energy of a phase according to composition at a fixed pressure and temperature.

According to relation [3.122], the derivative of the molar Gibbs energy with regard to the variable composition is therefore:

$$\frac{dG_m}{dx} = \mu_B - \mu_A \qquad [3.123]$$

This derivative is the slope of the tangent to the curve at composition point x. The equation of such a tangent is written:

$$y = mx + b$$

With, for the slope, according to relation [3.123]:

$$m = \mu_B - \mu_A \quad [3.124]$$

At point M, and composition point $x_{M,}$ the equation of the tangent will therefore be:

$$G_{m(M)} = \left(\mu_B - \mu_B^{(\alpha)}\right) x_M + b \quad [3.125]$$

By identification we find:

$$b = \mu_A \quad [3.126]$$

The new equation of the tangent being:

$$G_{m(M)} = \left(\mu_B - \mu_A\right) x + \mu_A \quad [3.127]$$

We see that this tangent intersects the axes $x = 0$ and $x = 1$ at two points P and Q such that:

$$BP = \mu_A \quad \text{and} \quad AQ = \mu_B \quad [3.128]$$

So, the tangent to the curve of the molar variable at composition point x intersects, on both vertical axes $x = 0$ and $x = 1$, two segments whose lengths are respectively the values of the corresponding partial molar variables for the chosen composition. This method of determining the chemical potentials is called the *tangent method*.

This property is true regardless of the molar variable chosen. This can therefore be applied when determining any partial molar variable of a binary solution, form the curve giving the integral molar variable according to the molar fraction.

In the case where one of the components is very dilute, for example when x tends towards zero (we tend towards the pure component B), the slope of the tangent to the curve can be written by clarifying the chemical potentials in relation [3.119] and an ideal behavior for B and an infinitely dilute behavior for A, we obtain:

$$b_0 = g_B^0 + RT \ln x - \mu_A^\infty + RT \ln(1-x) \qquad [3.129]$$

When x tends towards zero, this relation tends towards $-\infty$. We can show the same for $x = 1$ the slope of the tangent tends towards infinity. Thus, the tangents to the curve of the molar Gibbs energy at points $x = 0$ and $x = 1$ are vertical.

4

Statistics of Object Collections

Microscopic thermodynamics is a branch of statistical mechanics which are only concerned with thermodynamic applications. It is a way of describing a system from the entities that compose it; these entities may be called atoms, molecules, etc. It links macroscopic thermodynamic properties such as thermodynamic functions, thermodynamic coefficients, etc. to the microscopic properties of matter. This link is based on two hypotheses: that of the atomic structure of matter and that of the mean value.

4.1. The need to statistically process a system

4.1.1. *Collections, system description – Stirling's approximation*

The first hypothesis of microscopic thermodynamics is that of the atomic structure of matter and the application of laws of quantum mechanics.

If we consider, for example, a phase contained in volume V limited by boundaries called the system, usually a collection with a large number of objects is needed to describe this system, since these objects are so small that it is often necessary to apply the laws of quantum mechanics.

In fact, these objects are of very different natures: molecules, atoms, electrons, nuclei, protons, and neutrons with interaction bonds existing between different objects, i.e. very local or global interaction energies.

We therefore see that, from the description above, the system can be described as a number of objects but the definition of elementary objects is not required *a priori*.

Let us consider a container containing a certain amount of hydrogen gas. We can macroscopically define the system in at least three different ways:

– a collection of H_2 molecules, which under “normal conditions” do not interact with one another;

– a collection of hydrogen atoms with paired interactions;

– a collection of electrons and protons with complex interactions over short distances.

The experiment shows that the best choice, that whose problems caused by the microscopic approach are more easily resolved, is that whose objects in the collection have the fewest interactions between them; hence why in the previous example, we chose the molecular description.

If our container now contains a liquid mixture of alcohol and water, we would choose to describe the system by a double collection of water molecules and alcohol molecules. This description would not only cancel out the interactions between particles but would reduce these to three types over a short distance: interactions between two water molecules, between molecules of alcohol and between one molecule of water and one molecule of alcohol.

However, to determine certain properties of a metal, the best description often involves choosing a collection of electrons accompanying a collection of ions.

In the case where collections have a large number of objects such as molecules of one mole of a gas, i.e. 6.02×10^{23} molecules, we often find that the factorials of these large numbers, for which we can often use Stirling's approximation, can be written as follows for a large number n:

$$\ln n! \cong n(\ln n - 1) \cong n \ln n \qquad [4.1]$$

4.1.2. *Statistical description hypothesis*

Statistical thermodynamic reasoning is also based on a second hypothesis called the average value hypothesis, which states:

– having listed all the accessible states (or complexions) of a given collection, with each state being characterized by its own independent function, consistent with a given energy level, the value $\langle Q \rangle$ of the ensemble of property Q, likely to characterize the equilibrium state described, is the average of all accessible states of the values of Q by giving a unit statistical weight to each state, which can be written as follows:

$$\langle Q \rangle = \frac{\sum_{accessiblestate} Q}{\sum_{accessiblestate} 1} \qquad [4.2]$$

The denominator of the previous fraction represents the number of accessible states.

4.1.3. *The Boltzmann principle*

The state of the previously defined collection is constantly changing, such that over time the collection is distributed thereby its objects may find themselves in different states. The number of complexions is the number of distributions of objects between the different states that they are likely to take. Of all the possible distributions, there is one that is the maximum of the number of complexions. The Boltzmann principle states that the number of complexions corresponding to the most probably distribution type is almost equal to the total number of complexions and vice versa. The state of the collection therefore always corresponds to the maximum of complexions, which we call the *thermodynamic probability* or *dominant probability*.

This means that the distribution curve of the number of complexions between the different states shows a sharp peak corresponding to the maximum containing almost all the states.

4.2. Statistical effects of distinguishable non-quantum elements

First, we will consider a conventional approach. Initially we will only consider non-quantum objects, i.e. for which laws of conventional mechanics suffice to deduce their properties.

4.2.1. *Distribution law*

Each element has an energy ε_i and the number of elements with this energy is n_i. The total number of elements will be N, i.e.:

$$N = \sum_i n_i \qquad [4.3]$$

The total energy is therefore:

$$E = \sum_i n_i \varepsilon_i \qquad [4.4]$$

The number of complexions, i.e. configurations in the set of elements, is denoted Ω.

The mean energy of an element is $\langle \varepsilon \rangle$ and, by applying the relation [4.2], we obtain:

$$\langle \varepsilon \rangle = \frac{E}{N} \qquad [4.5]$$

The non-quantum elements composing the ensemble are distinguishable. Now, let us assume that each element only corresponds to one state, i.e. the statistical weight is equal to 1. The number of possible repartitions is the number of permutations with repetition (see Appendix 3, section A3.4.2.2) of N objects (here, elements) with n_0, n_1, ..., n_i that are similar but also separately identifiable, i.e. enumerating objects. We can change the $N!$ elements in N distinct ways; however, since the n_0 elements with energy ε_0 can be changed without changing the whole system, the results must be divided by $n_0!$; and also by $n_1!$, $n_2!$, $\dots n_i!$, etc. The number of complexions corresponding to our distribution is therefore:

$$\Omega = \frac{N!}{n_0! n_1! n_2! ... n_i! ...} \qquad [4.6]$$

Continuing this calculation, we will now assume that the number of elements is very high. Under these conditions, we can use Stirling's approximation [4.1] on numbers N, n_1, n_2, etc.

Relation [4.6] therefore becomes:

$$\ln \Omega \cong N(\ln N - 1) - \sum_i n_i (\ln n_i - 1) \cong N \ln N - \sum_i n_i \ln n_i \qquad [4.7]$$

We are now going to apply the Boltzmann principle by trying to find in which case the number of complexions Ω or $\ln\Omega$ is at its maximum.

We are therefore aiming to find the maximum of function $\ln\Omega$ given by expression [4.7], subject to two constraints [4.3] and [4.4]. For this, we will use the method of Lagrange multipliers (see Appendix 1) in this case with two constraints and multipliers, α for constraint [4.3] and β for constraint [4.4]. We therefore arrive at the following solution:

$$\ln n_i = -\alpha - \beta\varepsilon_i \qquad [4.8]$$

which can be written in the form:

$$n_i = \exp(-\alpha)\exp(-\beta\varepsilon_i) \qquad [4.9]$$

We have assumed that only one state existed for energy ε_i, when in fact there are several; several states g_i can correspond to each energy level ε_i. An element therefore has g_i chance more of occupying level ε_i than if there was only one corresponding state. The probability is therefore multiplied by g_i, and we say that this state is degenerate. g_i refers to the *statistical weight* or the *coefficient of degeneracy* or the *multiplicity* of the energy level ε_i. So, for the number of objects in state *i* it becomes:

$$n_i = g_i \exp(-\alpha)\exp(-\beta\varepsilon_i) \qquad [4.10]$$

Appendix 1 (see section A1.3) presents two methods to determine the values of the multipliers. We will use a specific value to determine α and a known phenomenon to determine β, where α is a dimensionless number whereas β is the inverse of an energy.

4.2.2. *Calculation of α*

We are going to determine α by using equality [4.3], i.e.:

$$N = \sum_i n_i = \exp(-\alpha)\sum_i g_i \exp(-\beta\varepsilon_i) \qquad [4.11]$$

And so:

$$\exp(-\alpha) = \frac{N}{\sum_i g_i \exp(-\beta\varepsilon_i)} \quad [4.12]$$

This value confirms that the coefficient α is in fact a number.

From expression [4.10], we obtain the Boltzmann distribution:

$$n_i = \frac{N g_i \exp(-\beta\varepsilon_i)}{\sum_i g_i \exp(-\beta\varepsilon_i)} \quad [4.13]$$

This distribution can also be expressed in terms of the fraction of the number of elements in state i:

$$p_i = \frac{n_i}{N} = \frac{g_i \exp(-\beta\varepsilon_i)}{\sum_i g_i \exp(-\beta\varepsilon_i)} \quad [4.14]$$

Let us now determine coefficient β.

4.2.3. *Determining coefficient β*

We are going to use the second method mentioned in Appendix 1 (see section A1.3), in which we use a Lagrange multiplier coefficient to determine coefficient β found in different distributions.

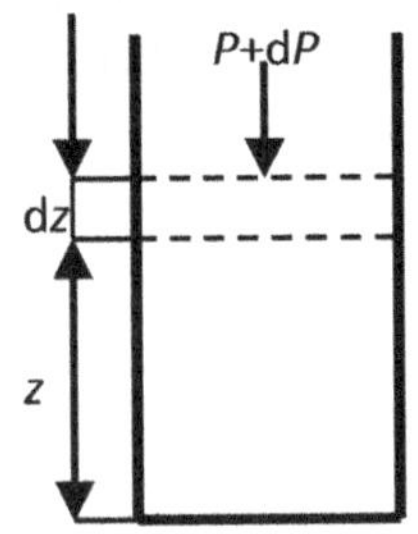

Figure 4.1. *Height distribution of gas molecules under the action of gravity*

Here, we consider an isothermal column of a unit section, containing a gas with a molar mass of M and split this column into small elements. Each element has a height dz and is situated in the z section (Figure 4.1).

The energy, here the energy of gravity, corresponds to an element is ε_i, such that:

$$\varepsilon_i = mgz \quad [4.15]$$

Here, the energy levels are very close together and their discontinuous distribution can be replaced by a continuous one and so that that the statistical weight of states corresponding to the layer with a thickness of dz is:

$$g_i = K\,\mathrm{d}\,z \quad [4.16]$$

K is a constant (for a given gas at a given temperature). The number of molecules $\delta\, n_i$, with energies between mgz and $mg(z+\mathrm{d}\,z)$, is, by applying expression [4.10]:

$$\delta\, n_i = K\,\mathrm{d}\,z \exp(-\alpha)\exp(-\beta mgz) \quad [4.17]$$

Let ρ be the number of molecules per unit volume (molecular concentration). Since the column has a unit section, we obtain:

$$\rho = \frac{\delta n_i}{\mathrm{d}z} \quad [4.18]$$

Using expression [4.17], this concentration becomes:

$$\rho = \exp(-\alpha)K\exp(-\beta mgz) = \rho_0 \exp(-\beta mgz) \quad [4.19]$$

ρ_0 is constant for a given gas at a given temperature.

Relation [4.19] gives the result deduced from the Maxwell–Boltzmann distribution.

We can now proceed to a direct calculation. The difference in pressure (dP) between the upper and lower part of element dz corresponds to the weight of this element, with M as the molar mass and C the molar concentration of the gas:

$$-\mathrm{d}P = CMgz \qquad [4.20]$$

If the gas is ideal, then:

$$P = CRT \qquad [4.21]$$

i.e.:

$$\mathrm{d}P = \mathrm{R}T\,\mathrm{d}C \qquad [4.22]$$

Using equation [4.21] it becomes:

$$\frac{\mathrm{d}C}{C} = -\frac{Mg}{\mathrm{R}T}\mathrm{d}z \qquad [4.23]$$

The integration of this expression gives:

$$C = C_0 \exp\left(-\frac{Mgz}{\mathrm{R}T}\right) \qquad [4.24]$$

If m is the mass of the molecule of gas, $\mathrm{k_B}$ and $\mathrm{N_a}$ are the Boltzmann constant and the Avogadro constant, respectively, such that:

$$M = m\mathrm{N_a} \qquad [4.25]$$

and

$$\mathrm{k_B} = \frac{\mathrm{R}}{\mathrm{N_a}} \qquad [4.26]$$

Equation [4.24] becomes:

$$C = C_0 \exp\left(-\frac{mgz}{\mathrm{k_B}T}\right) \qquad [4.27]$$

By definition we obtain:

$$\rho = C\mathrm{N_a} \qquad [4.28]$$

And finally:

$$\rho = \rho_0 \exp\left(-\frac{mgz}{\mathrm{k_B}T}\right) \qquad [4.29]$$

With both relations [4.19] and [4.29] being identical, by identifying β we obtain:

$$\beta = \frac{1}{\mathrm{k_B}T} \qquad [4.30]$$

This variable is homogeneous to the inverse of an energy and therefore depends only on temperature.

4.2.4. *Energy input to a system*

The total energy of a group of particles of the same species can be written:

$$U - U(0) = E = \sum_i n_i \varepsilon_i \qquad [4.31]$$

The variation in this energy, during a reversible transformation, will have the form:

$$\mathrm{d}U = \sum_i n_i \,\mathrm{d}\varepsilon_i + \sum_i \varepsilon_i \,\mathrm{d}n_i \qquad [4.32]$$

Note that a modification in the total energy of a group is due to two factors:

– either a modification in the element energies (from ε_i to $\varepsilon_i + \mathrm{d}\varepsilon_i$) which corresponds to the first term in the previous relation;

– or a change in the population at a given level (change from n_i to $n_i + \mathrm{d}n_i$) which corresponds to the second term in the previous relation.

When a system is heated, the energy levels ε_i are unchanged but their population n_i is modified. When the system exchanges work, this time the energies are modified. So, in expression [4.32], during heating, only the second term is modified.

4.2.5. *The Boltzmann principle for entropy*

As we have just seen, when a system is heated at a constant volume, the energy levels do not vary; therefore, without any modification other than heating, the variation in energy of the ensemble will be given by the heat involved in a reversible transformation, which corresponds to the first term in the second part of equation [4.32]:

$$\mathrm{d}U = \sum_i \varepsilon_i \,\mathrm{d}n_i = dQ_{\mathrm{rev}} = TdS \qquad [4.33]$$

From this, we deduce the variation in entropy of the ensemble in a reversible transformation:

$$\mathrm{d}S = \frac{1}{T}\sum_i \varepsilon_j \,\mathrm{d}n_i = \mathrm{k_B}\beta \sum_j \varepsilon_i \,\mathrm{d}n_i \qquad [4.34]$$

By combining relations [4.7] and [4.8], the variation in the most probable configuration is:

$$\frac{\partial \ln \Omega}{\partial n_i} + \alpha + \beta\varepsilon_i = 0 \qquad [4.35]$$

Then, we deduce:

$$\beta\varepsilon_i = -\frac{\partial \ln \Omega}{\partial n_i} - \alpha \qquad [4.36]$$

Using relation [4.34], it becomes:

$$\mathrm{d}S = \mathrm{k_B}\sum_i \left(\frac{\partial \ln \Omega}{\partial n_i}\right) \mathrm{d}n_i + \mathrm{k_B}\alpha \sum_i \mathrm{d}n_i \qquad [4.37]$$

As the amount of matter is fixed in each element, we obtain:

$$\sum_i \mathrm{d}n_i = 0 \qquad [4.38]$$

And therefore:

$$\mathrm{d}S = \mathrm{k_B}\sum_i \left(\frac{\partial \ln \Omega}{\partial n_i}\right)\mathrm{d}n_i = \mathrm{k_B}\,\mathrm{d}\ln\Omega \qquad [4.39]$$

By integration between a state with a unit complexion and the actual state, we obtain:

$$S = S_0 + \mathrm{k_B}\ln\Omega \qquad [4.40]$$

If T decreases, according to Planck's law, the entropy decreases and if T tends towards zero, there is only one configuration, that corresponding to the fundamental state, i.e. $\Omega=1$ and $\ln\Omega=0$; the term S_0 is therefore zero, which is consistent with the third principle of thermodynamics and the entropy of the system becomes:

$$S = \mathrm{k_B}\ln\Omega \qquad [4.41]$$

This is the Boltzmann formula. This formula is featured, as an epitaph, on Boltzmann's tomb.

4.3. The quantum description and space of phases

4.3.1. *Wave functions and energy levels*

Now let us take the example where our collection of particles requires, in order to be described, the use of quantum mechanics and a stationary state that satisfies Schrödinger's equation of the system. We know that to solve this equation, we must determine the energy values E, called eigenvalues, that lead to acceptable wave functions[1], which are the solution to the previous equation, called eigenfunctions. In fact, the energy values found can be quite close and numerous with the need to consider an energy interval

1 The term "acceptable" refers to the fact that wave functions must have certain properties: to be hermitic, of an integrable square, etc.

between E and E + dE, called an energy level. This level may have several corresponding eigenfunctions. This level will therefore correspond to a state i containing g_i solutions, which has a statistical weight g_i or that degenerates g_i times.

4.3.2. *Space of phases: discernibility of objects and states*

Let us consider a collection of N identical "quantum" objects each with three degrees of freedom (e.g. one punctual gas molecule). To define the movement of these objects, we must first define the three spatial coordinates and the three quantities of movement (or three velocity components), i.e. a total of six coordinates in a space with six dimensions. In this space, called the quantum space, an object is defined by a point. For N objects, we use an hyperspace with 6N dimensions, called the Gibbs' space of phase. In this hyperspace, it is a system state which is represented by a point. Each ensemble defined constitutes a complexion of the system.

In fact, according to the Heisenberg's uncertainty principle, we cannot accurately give the state of a particle by a point in a six dimensional space because the uncertainty in this space gives a hypercube with a side equal to the square root of Planck's constant h. This uncertainty prevents us from accurately defining the state of the system as the previous hypercube leads to a hyper-parallelepiped called a Heisenberg region in the space of phases. Over time, the system continuously evolves and we are unable to distinguish two objects that are simultaneously in the same hypercube, hence we cannot know if an object that enters the hypercube is the same as that which left passing to the next complexion. Two objects belonging to the same hypercube in the space with six dimensions belong to the same Heisenberg region in the space of phases. Supposedly, the particles are indiscernible and we are unable to completely follow their evolution.

Of course, if the objects concerned are not "quantum" objects, Heisenberg's principle no longer applies and the evolutions can be followed from one state to the next; the objects are supposedly discernible. We will find an example of this in our study of a canonical ensemble.

4.3.3. *Localization and non-localization of objects*

Let us consider an ensemble composed of quantum objects such as molecules, atoms, punctual defects of a solid and of nuclei, protons, electrons and neutrons that we will group under the name "molecules" to specify that these are quantum objects. If each of these molecules limits displacements in a specific region of space such that two molecules never belong to the same Heisenberg region in the quantum space, these molecules are said to be isolated. This is due to the fact that they become discernible since none of these molecules can be continuously tracked over time. Therefore, a collection of isolated molecules is a collection of discernible particles.

If we consider the three states of matter, the molecules of a crystallized solid constitute localized, and hence discernible, molecules. However, an ensemble of molecules of a gas or liquid constitutes ensembles of non-isolated, and hence indiscernible, molecules.

4.4. Statistical effect of localized quantum objects

Let us consider an ensemble of N localized, and hence discernible, molecules, such that its volume V, total energy E and the number N remain constant. In a given state, each molecule has an energy ε_i, with the states being degenerated g_i times. We often call this collection a *microcanonical* ensemble. We can therefore write:

$$N = \sum_i n_i \tag{4.42}$$

and

$$E = \sum_i n_i \varepsilon_i \tag{4.43}$$

We will find that, in the same case as in section 4.2, the number of ways our N elements can be arranged in the space of phases is given by an expression in the same form as inequality [4.6] where we can now assume that each energy level has several corresponding states g_i for ε_i. The probability is therefore multiplied by g_i, the same for each element, and for all elements, factor $g_i^{n_i}$. In total, the number of complexions is:

$$\Omega = \frac{N!\prod_i g_i^{n_i}}{\prod_i n_i!} \qquad [4.44]$$

This is the same Maxwell-Boltzmann statistic and we can therefore write:

$$n_i = g_i \exp(-\alpha)\exp(-\beta\varepsilon_i) \qquad [4.45]$$

and

$$p_i = \frac{n_i}{N} = \frac{g_i \exp(-\beta\varepsilon_i)}{\sum_i g_i \exp(-\beta\varepsilon_i)} \qquad [4.46]$$

The coefficient β has the same value as above (see section 4.7) and the coefficient α is deduced in a similar way to equation [4.12] by:

$$\exp(-\alpha) = \frac{N}{\sum_i g_i \exp(-\beta\varepsilon_i)} \qquad [4.47]$$

and the law of distribution of our discernible molecules will be:

$$p_i = \frac{n_i}{N} = \frac{g_i \exp(-\beta\varepsilon_i)}{\sum_i g_i \exp(-\beta\varepsilon_i)} \qquad [4.48]$$

NOTE.– The sum that appears in the denominator in expressions [4.47] and [4.48] is called the *molecular partition function*, defined by:

$$z = \sum_i g_i \exp(-\beta\varepsilon_i) \qquad [4.49]$$

4.5. Collections of non-localized quantum objects

In this section we will consider collections of indiscernible, and hence non-localized, objects.

4.5.1. *Eigen symmetrical and antisymmetric functions of non-localized objects*

Let us consider a collection of N objects with weak interactions between one another. Schrödinger's equation for the ensemble of the collection will be:

$$\left[\mathrm{H}\left(x_1 - \frac{i\mathrm{h}}{2\pi}\frac{\partial}{\partial x_1}\right) + \mathrm{H}\left(x_2 - \frac{i\mathrm{h}}{2\pi}\frac{\partial}{\partial x_2}\right) + \ldots - E\right]\Psi = 0 \qquad [4.50]$$

The eigenvalue is in the following form:

$$E = \varepsilon_\sigma + \varepsilon_\tau + \ldots + \varepsilon_\omega \qquad [4.51]$$

Two indices $\sigma, \tau \ldots \omega$ are never equal.

This ensemble corresponds to a series of $N!$ eigenfunctions obtained by permutation (see Appendix 3, section A3.4.2) of objects 1, 2, … N between the indices of eigenvalues $\sigma, \tau, \ldots \omega$. An example of such a function would be:

$$\Psi = \psi_\sigma(x_1) + \psi_\tau(x_2) + \ldots + \psi_\omega(x_N) \qquad [4.52]$$

For systems respectively degenerated $g_\sigma, g_\tau, \ldots, g_\omega$ times, the total number of eigenfunctions will be $N!\, g_\sigma g_\tau \ldots g_\omega$.

Finally, if we consider the complete general eigenvalue:

$$E = n_\sigma \varepsilon_\sigma + n_\tau \varepsilon_\tau + \ldots + n_\omega \varepsilon_\omega \qquad [4.53]$$

where the individual eigen-solutions are grouped into $n_\sigma, n_\tau \ldots n_\omega$ ways, such that:

$$N = n_\sigma + n_\tau + \ldots + n_\omega \qquad [4.54]$$

A series of $N!/n_\sigma! + n_\tau! + \ldots + n_\omega!$ eigenfunctions obtained by permutations exists, an example of which would be:

$$\Psi=\psi_\sigma(x_1)+...+\psi_\sigma(x_{n_\sigma})+\psi_\tau(x_{n_\sigma+1})+...+\psi_\tau(x_{n_\sigma+n_\tau})..\psi_\omega(x_N) \quad [4.55]$$

The total number of degenerated eigenfunctions is therefore:

$$\frac{N!g_\sigma^{n_\sigma} g_\tau^{n_\tau} ...g_\omega^{n_\omega}}{n_\sigma!+n_\tau!+...+n_\omega!} \quad [4.56]$$

Any linear combination of these eigenfunctions is also the eigenfunction for the same value of energy, such that linear combinations can be formed in a number equal to the number of independent eigenfunctions and used in their place.

We show that eigenfunctions, such as those given by relation [4.55], of a series of similar systems, after reorganization into linear combinations, are necessarily grouped into a certain number of groups, *A*, *B*, *S* etc., defined from their properties of symmetry. These groups possess an interesting property: no force and no symmetric interaction, with regard to the coordinates of similar systems, can modify the assembly of an eigenfunction of group *A*; for example, in an eigenfunction of another group *B*. So, if an assembly is initially represented by an eigenfunction of group *A,* it will always be defined by the same and only these states will be accessible.

From groups *A, B, S,* we distinguish:

– the symmetrical group for which the eigenfunctions are symmetrical throughout the whole system and called group *S*;

– the antisymmetric group of group *A* whose wave function takes the form of the determinant:

$$\begin{vmatrix} \psi_\sigma(x_1) & \psi_\sigma(x_2) & \cdots\cdots & \psi_\sigma(x_N) \\ \psi_\tau(x_1) & \psi_\tau(x_2) & \cdots\cdots & \psi_\tau(x_N) \\ \cdots & \cdots & \cdots & \cdots \\ \psi_\omega(x_1) & \psi_\omega(x_2) & \cdots\cdots & \psi_\omega(x_N) \end{vmatrix} \quad [4.57]$$

For group *S,* we use the same determinant but with minus signs instead.

We find that for symmetrical wave functions (group S), the objects can occupy any state in any number, whereas for antisymmetric wave functions (group A), there can only be one object in a given state (e.g. electrons according to the Pauli exclusion principle). So, the number of complexions will be different in both cases and each will correspond to a particular distribution:

– for type S wave functions, Bose-Einstein statistics will be applied;

– for type A wave functions, Fermi–Dirac statistics will be applied.

In practice, Bose-Einstein statistics concern particles with zero or full spin: photons, α-particles; whereas Fermi-Dirac statistics concern particles with half-integer spin: electrons, protons, neutrons.

As far as molecules and ions are concerned, they are composed of p protons, n neutrons and l electrons:

– if $n + p + l$ is an even number, that is to say if the sum of spins is an integer, the system is symmetrical and Bose-Einstein statistics is used;

– if $n + p + l$ is an odd number, that is to say if the sum of the spins is a half-integer, the system is antisymmetric and Fermi–Dirac statistics is used.

As these systems do not change symmetry, a system can therefore be considered to be composed of either a mixture of protons, neutrons and electrons, or a mixture of atoms with the same symmetry, or an ensemble of molecules still with the same symmetry, or a group of molecules with the same symmetry. The ensemble of objects that corresponds to the minimum interactions between objects is always chosen.

4.5.2. *Statistics of non-localized elements with symmetrical wave functions*

We saw that for symmetrical wave functions, Bose-Einstein statistics is applied.

In this case, we can only distribute N objects into the groups n_1, n_2, … n_i in one way since they are indiscernible. Then, the n_i identical objects are distributed between g_i energy states and represented by g_i cases (Figure 4.2) and are therefore separated by $g_i - 1$ bars. We can swap n_i objects and $g_i - 1$ bars, i.e. $n_i + g_i - 1$ objects, and consequently ensure $(n_i + g_i - 1)!$ permutations.

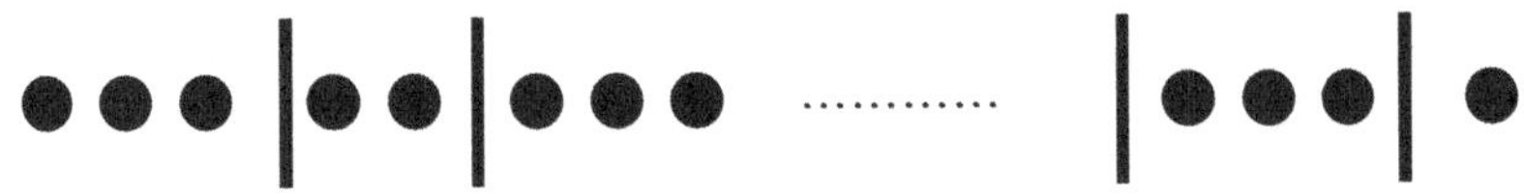

Figure 4.2. *Arrangement of n_i indiscernible balls between g_i cases*

The indiscernibility of objects and bars remains to be discussed. If a permutation of objects is carried out when the bars and the population of cases are fixed, nothing changes as they are indiscernible. As there are $n_i!$ corresponding permutations, the number of permutations calculated above is $n_i!$ times greater. It is $(g_i - 1)!$ times larger for the bar permutation. The number of permutations of n_i objects in g_i cases or n_i particles in g_i states is therefore:

$$\frac{(n_i + g_i - 1)!}{n_i!(g_i - 1)!} \qquad [4.58]$$

The total number of complexions is therefore:

$$\Omega_{BE} = \prod_i \frac{(n_i + g_i - 1)!}{n_i!(g_i - 1)!} \qquad [4.59]$$

We will now apply Stirling's formula [4.1] to relation [4.59], which requires that the numbers n_i and g_i be very large, g_i in particular. This is true in the case of molecules when the temperature reaches 1K. At very low temperatures, if there is only one simple level, we would obtain $g_i = 1$ and the same for levels immediately above; however, above 1K, the number of molecules in the lower levels is so small that g_i is considered to be very large.

We will also omit 1 with regards to n_i and g_i, and, by passing through the logarithms, we obtain:

$$\ln \Omega_{BE} \cong \sum_i \left[(n_i + g_i) \ln (n_i + g_i) - n_i \ln n_i - g_i \ln g_i \right] \qquad [4.60]$$

By varying the number of molecules in quantity δn_i, and by differentiation, we obtain:

$$\delta \ln \Omega_{\text{BE}} = \sum_i \left[\ln\left(\frac{n_i + g_i}{n_i} \right) \right] \delta n_i \qquad [4.61]$$

By proceeding as we have done previously, we will obtain three relations:

$$\sum_i \left[\ln\left(\frac{n_i + g_i}{n_i} \right) \right] \delta n_i = 0 \qquad [4.62]$$

$$\sum \delta n_i = 0 \qquad [4.63]$$

$$\sum \varepsilon_i \, \delta n_i = 0 \qquad [4.64]$$

Again, we will apply the Lagrange multiplier method with two conditions (see Appendix 1) and therefore two coefficients. Let α be the coefficient relative to the number of molecules and β the coefficient relative to energy. We obtain the solution:

$$(n_i)_{BE} = \frac{g_i \exp(-\alpha - \beta\varepsilon_i)}{1 - \exp(-\alpha - \beta\varepsilon_i)} \qquad [4.65]$$

We will cover this later on (see sections 4.7 and 4.8) on values of α and β.

4.5.3. *Statistics of non-localized elements with an asymmetric function*

We have seen that in the case of asymmetrical wave functions Fermi–Dirac statistics is applied.

In this case, the objects can only be distributed in one way between the levels ε_1, ε_2, … ε_i. Then, the identical n_i particles are distributed between g_i states, but as in Fermi–Dirac statistics, there can only be a maximum of one object per case, which means that the number of cases must be greater than the number of objects and therefore:

$$g_i > n_i \qquad [4.66]$$

We have g_i possibilities to place the first object, $g_i - 1$ possibilities for the second and $g_i - n_i + 1$ for the n_i^{th}, i.e. a total number of possibilities:

$$g_i(g_i-1)(g_i-2)...(g_i-n_i-1)=\frac{g_i!}{(g_i-n_i)!} \qquad [4.67]$$

Having counted the objects, as they are indiscernible, we can swap them in any form without changing anything. We must therefore divide the previous number by $n_i!$, i.e. for the distribution of n_i objects in g_i cases:

$$\frac{g_i!}{(g_i-n_i)!n_i!} \qquad [4.68]$$

We must restart the process for each species i, therefore the total number of complexions will be:

$$\Omega_{FD}=\prod_i \frac{g_i!}{(g_i-n_i)!n_i!} \qquad [4.69]$$

To discern the distribution law, we will proceed in exactly the same way as above to obtain:

$$(n_i)_{FD}=\frac{g_i\exp(-\alpha-\beta\varepsilon_i)}{1+\exp(-\alpha-\beta\varepsilon_i)} \qquad [4.70]$$

NOTE.– We can summarize the three distribution laws of quantum objects [4.45], [4.65] and [4.70] in the form:

$$n_i=\frac{g_i\exp(-\alpha-\beta\varepsilon_i)}{1+\varepsilon\exp(-\alpha-\beta\varepsilon_i)} \qquad [4.71]$$

agreeing that ε has the value:

– $\varepsilon = 0$ in the Maxwell-Boltzmann distribution for localized particles;

– $\varepsilon = -1$ in Bose-Einstein statistics for non-localized integer spin particles;

– $\varepsilon = +1$ in Fermi–Dirac statistics for non-localized half-integer spin particles.

4.5.4. *Classical limiting case*

The three quantum statistics combine, if, in fact, we satisfy the condition:

$$\exp(-\alpha - \beta\varepsilon_i) << 1 \qquad [4.72]$$

Now, β and ε_i are both positive and the previous condition is reduced to:

$$\exp(-\alpha) << 1 \qquad [4.73]$$

therefore α must be very large.

Under these conditions, the three laws combine to form:

$$n_i = g_i \exp(-\alpha - \beta\varepsilon_i) \qquad [4.74]$$

This is the limit known as the *classical limiting case*, which maintains the Maxwell–Boltzmann distribution formula.

If expression [4.73] is satisfied such that $N \ll z$, given by relation [4.49] and since β and ε_i are positive, then we obtain:

$$z < \sum_i g_i \qquad [4.75]$$

And finally:

$$N << \sum_i g_i \qquad [4.76]$$

The classical limiting case occurs when the number of particles is far lower than the number of possible states. In practice, this occurs with gas molecules if the temperature is not too low.

4.6. Systems composed of different particles without interactions

Now, let us assume that our system is formed of a double collection of molecules of two different species A and B, and that these molecules are completely independent.

For example, this case will occur if the system is composed of two constituents belonging to two different phases. However, this can also be the case for two constituents belonging to the same phase without interaction between the molecules.

The total number of complexions in our system is given by the product of the number of complexions of both sub-ensembles:

$$\Omega_{A,B} = \Omega_A \Omega_B \qquad [4.77]$$

The appropriate statistic is applied separately to each sub-ensemble, which each behave as if they were independent.

4.7. Unicity of coefficient β

Whatever the statistic chosen, both for microcanonical ensembles and for quantum systems, the Lagrange multiplier method introduces two multiplier coefficients, which we have not characterized for the different statistics: β, relative to energy, and α, relative to the number of objects.

For coefficient α, it is determined every time by writing that the number of particles is equal to N. In the case of a mixture with several constituents, A, B, C etc., by number of molecules, N_A, N_B, N_C etc., each constituent will have an α value: α_A, α_B, α_C, etc.

For coefficient β, we have implicitly assumed that it will have the same value for all the statistics found. We are going to describe this property.

Let us consider a mixture of two families of particles, 1 and 2, without interaction obeying two different quantum statistics. The total number of complexions is, according to equality [4.77], the product of the complexions of two sub-systems:

$$\Omega = \Omega_1(n_i)\Omega_2(n_k) \qquad [4.78]$$

By switching to logarithms, it becomes:

$$\ln \Omega = \ln \Omega_1(n_i) + \ln \Omega_2(n_k) \qquad [4.79]$$

Ω must be maximum, thus:

$$\mathrm{d}\ln\Omega = 0 \qquad [4.80]$$

which is written:

$$\sum_{i=1}^{N_1} \frac{\partial \ln \Omega_1}{\partial n_i} \mathrm{d} n_i + \sum_{k=1}^{N_2} \frac{\partial \ln \Omega_2}{\partial n_k} \mathrm{d} n_k = 0 \qquad [4.81]$$

This time we have three constraints given by the relations [4.82], [4.83] and [4.84]; two constraints on the numbers N_1 and N_2 of particles in two sub-systems and one constraint on the total energy of the system. The three parameters α_1, α_2 and β give:

$$\sum_{i=1}^{N_1} n_i = N_1 \text{ i.e. } \sum_{i=1}^{N_1} \mathrm{d} n_i = 0 \qquad [4.82]$$

$$\sum_{k=1}^{N_2} n_k = N_2 \text{ i.e. } \sum_{k=1}^{N_2} \mathrm{d} n_k = 0 \qquad [4.83]$$

$$\sum_{i=1}^{N_1} \varepsilon_i n_i + \sum_{k=1}^{N_2} \varepsilon_k n_k = U \text{ i.e. } \sum_{i=1}^{N_1} \varepsilon_i \, \mathrm{d} n_i + \sum_{k=1}^{N_2} \varepsilon_k \, \mathrm{d} n_k = 0 \qquad [4.84]$$

By applying the Lagrange multiplier method (Appendix 1) we deduce:

$$\left[\sum_{i=1}^{N_1} \frac{\partial \ln \Omega_1}{\partial n_i} + \alpha_1 + \beta \sum_i \varepsilon_i \right] \mathrm{d} n_i + \left[\sum_{k=1}^{N_2} \frac{\partial \ln \Omega_2}{\partial n_k} + \alpha_2 + \beta \sum_k \varepsilon_k \right] \mathrm{d} n_k = 0$$

[4.85]

With the numbers n_i being independent from numbers n_k we can simultaneously deduce the following:

$$\left[\sum_{i=1}^{N_1} \frac{\partial \ln \Omega_1}{\partial n_i} + \alpha_1 + \beta \sum_i \varepsilon_i \right] \mathrm{d} n_i = 0 \qquad [4.86]$$

$$\left[\sum_{k=1}^{N_2} \frac{\partial \ln \Omega_2}{\partial n_k} + \alpha_2 + \beta \sum_k \varepsilon_k\right] \mathrm{d} n_k = 0 \qquad [4.87]$$

The numbers n_i are independent of one another:

$$\left[\sum_{i=1}^{N_1} \frac{\partial \ln \Omega_1}{\partial n_i} + \alpha_1 + \beta \sum_i \varepsilon_i\right] = 0 \qquad [4.88]$$

Similarly, the numbers n_k are also independent of one another:

$$\left[\sum_{k=1}^{N_2} \frac{\partial \ln \Omega_2}{\partial n_k} + \alpha_2 + \beta \sum_k \varepsilon_k\right] = 0 \qquad [4.89]$$

We see that if we solve these two equations, each in its own statistic, β would have the same value.

So, the value of coefficient β is the same regardless of the statistic applied and is therefore given by expression [4.30].

4.8. Determining coefficient α in quantum statistics

Whereas coefficient β is universal, coefficient α, relative to quantities of matter, will be specific to each species of element and each statistic used. In order to calculate coefficient α from the total number of molecules:

$$\sum_{i=1}^{N_1} n_i = N \qquad [4.90]$$

When calculating this in Bose–Einstein (relation [4.65]) and Fermi–Dirac (relation [4.70]) statistics, it is impossible to determine this coefficient. We must therefore settle for a limiting development in N/z in the form:

$$\exp(-\alpha_{\mathrm{A}}) = a_0 + a_1 \frac{N_{\mathrm{A}}}{z_{\mathrm{A}}} + a_2 \left(\frac{N_{\mathrm{A}}}{z_{\mathrm{A}}}\right)^2 + a_3 \left(\frac{N_{\mathrm{A}}}{z_{\mathrm{A}}}\right)^3 + \ldots \qquad [4.91]$$

Laborious calculations show that the a_i coefficients of this development are:

$$a_0 = 0 \; ; \; a_1 = 1 \; ; \; a_2 = \pm\frac{1}{2^{3/2}} \; ; \; a_3 = \frac{1}{4} - \frac{1}{3^{3/2}} \; ; .. \qquad [4.92]$$

In coefficient a_2, the "+" sign is applied with Fermi Dirac statistics and the "–" sign with Bose–Einstein statistics.

If we place this development in the most probable distribution, it becomes:

– in Bose–Einstein statistics with $n_i >> 0$:

$$n_{i(\mathrm{A})} = \frac{g_{i(\mathrm{A})}}{\left[\left(\frac{N_\mathrm{A}}{z_\mathrm{A}}\right) - \frac{1}{2^{3/2}}\left(\frac{N_\mathrm{A}}{z_\mathrm{A}}\right)^2 + \left(\frac{1}{4} - \frac{1}{3^{3/2}}\right)\left(\frac{N_\mathrm{A}}{z_\mathrm{A}}\right)^3 + \ldots\right]^{-1} \exp\frac{\varepsilon_i}{kT} - 1}$$

[4.93]

– in Fermi–Dirac statistics with the conditions $g_i >> n_i >> 0$:

$$n_{i(\mathrm{A})} = \frac{g_{i(\mathrm{A})}}{\left[\left(\frac{N_\mathrm{A}}{z_\mathrm{A}}\right) + \frac{1}{2^{3/2}}\left(\frac{N_\mathrm{A}}{z_\mathrm{A}}\right)^2 + \left(\frac{1}{4} - \frac{1}{3^{3/2}}\right)\left(\frac{N_\mathrm{A}}{z_\mathrm{A}}\right)^3 + \ldots\right]^{-1} \exp\frac{\varepsilon_i}{kT} + 1}$$

[4.94]

In the classical limiting case, we know that the three statistics merge and are given by relation [4.74]. If α is large then we are limited to the first term of the development [4.91], which gives:

$$\exp(-\alpha_\mathrm{A}) \cong \frac{N_\mathrm{A}}{z_\mathrm{A}} \qquad [4.95]$$

We have therefore defined three statistics used for well-defined collections of particles. These three statistics can, under certain conditions, be reduced to just one, the *classical limiting case.* We will use these statistics to study collections of *molecules*, in the most general sense, represent phases.

5

Canonical Ensembles and Thermodynamic Functions

In Chapter 4, we saw statistical methods to process collections of quantum or non-quantum objects. Now, we will assume that not only the entropy as the Boltzmann theorem reveals, but all the thermodynamic properties of a constituent can be accurately obtained by considering that this constituent is in fact a collection of many identical quantum objects. Similarly, a mixture with several constituents will be considered as many collections of quantum objects. So, we thus move onto the concept of non-quantum ensembles with an infinite number of system reproductions under well-defined conditions: a canonical ensemble.

5.1. An ensemble

We are going to specify the meaning of the term "ensemble" in the field of statistical thermodynamics, whose definition will be a bit more precise in this context than in everyday language.

An ensemble is an identical replica of a system. This replica is imaginary; the ensemble does not physically exist, which allows us to readily choose the number of elements N_E, as many as one wishes, or if needed, to make it tend towards infinity. The ensemble has the energy $\mathrm{E_E}$ and contains a certain number of elements n_j whose energy is E_j. The ensemble could be considered a collection of non-quantum objects, or elements, and we can describe the configurations or complexions of the ensemble.

There are several types of ensembles used in statistical thermodynamics (see section 5.6). The ensemble that we will mainly focus on in this chapter is called a "canonical ensemble".

5.2. Canonical ensemble

Canonical ensembles are widely used; they are particularly well-adapted to the microscopic description of solid or fluid phases, which must be studied in detail.

5.2.1. *Description of a canonical ensemble*

We call a *canonical ensemble* an ensemble (in the previous sense) composed of replicas of the studied system; each element is closed, therefore the number of molecules N is identical in all elements of the ensemble. This number is also constant in the same element (no exchange of matter between elements and between elements and the external environment). The volume V is the same for all elements. The elements are in thermal contact with one another and can therefore exchange energy; their temperature is identical T. The total energy of the canonical ensemble will be E_C. This energy is constant since the ensemble is isolated from the external environment.

Each element has an energy E_j and the number of elements with this energy is n_j. The total number of elements of the canonical ensemble will be N_C, such that:

$$N_C = \sum_j n_j \tag{5.1}$$

The total energy of the canonical ensemble is therefore:

$$E_C = \sum_j n_j E_j \tag{5.2}$$

The number of complexions, i.e. configurations of the ensemble of elements in the canonical ensemble, will be Ω_C.

The mean energy of an element in the canonical ensemble is $\langle E \rangle$ and, by applying the definition of relation [5.2], we obtain:

$$\langle E \rangle = \frac{E_C}{N_C} \quad [5.3]$$

or

$$\langle E \rangle = \frac{\sum_j n_j E_j}{\sum_j n_j} \quad [5.4]$$

An element of the canonical ensemble contains N objects. The energy of object i is ε_i and the number of objects with energy ε_i is n_i, such that:

$$N = \sum_i n_i \quad [5.5]$$

Usually, the index j will be assigned to elements of the canonical ensemble, with index i being reserved for objects of the system, i.e. objects of an element in the canonical ensemble.

NOTE.– There is no relation between the number of objects in an element N of the canonical ensemble, which is fixed by the system, and the number of systems in the canonical ensemble N_C, which is arbitrary.

5.2.2. *Law of distribution in a canonical ensemble*

The canonical ensemble is composed of non-quantum elements, which are distinguishable. Maxwell-Boltzmann statistics can therefore be applied, giving relation [5.13].

We therefore obtain the Maxwell Boltzmann distribution:

$$n_j = \frac{N_C \exp\left(-\beta E_j\right)}{\sum_j \exp\left(-\beta E_j\right)} \quad [5.6]$$

This distribution can also be expressed as a fraction of the number of elements in state j:

$$p_j = \frac{n_j}{N_C} = \frac{\exp(-\beta E_j)}{\sum_j \exp(-\beta E_j)} \quad [5.7]$$

5.2.3. *Canonical partition function*

When the statistical results are used in the calculation of thermodynamic data, the following equation often appears, as in relations [5.6] and [5.7]:

$$\sum_j \exp(-\beta E_j)$$

which Fowler named the *canonical partition function*, whereas Planck called it the *canonical state equation.*

We will continue to use that of Fowler and represent it by Z_C, i.e.:

$$Z_C = \sum_j \exp(-\beta E_j) \quad [5.8]$$

and the Maxwell–Boltzmann law of distribution becomes:

$$p_j = \frac{n_j}{N_C} = \frac{\exp(-\beta E_j)}{Z_C} \quad [5.9]$$

with:

$$\beta = \frac{1}{\mathrm{k_B} T} \quad [5.10]$$

Hence, the Maxwell–Boltzmann law of distribution becomes:

$$p_j = \frac{n_j}{N_C} = \frac{g_j \exp\left(-\frac{E_j}{\mathrm{k_B}T}\right)}{Z_C} \qquad [5.11]$$

5.3. Molecular partition functions and canonical partition functions

Quantum mechanics helps determine the energies of molecules and the molecular partition functions, as we will see in Chapter 6, can be calculated from the characteristics of molecules (mass, moments of inertia, vibrational frequencies, etc.). The aim of this section is to determine the value of the canonical partition functions, since this is useful to link them to molecular partition functions.

5.3.1. *Canonical partition functions for ensembles of discernable molecules*

Firstly, let us consider an ensemble of localized, i.e. discernable, molecules. We can write that the energy of an element j of the canonical ensemble is the sum of the energies of the molecules of each element, i.e.:

$$E_j = \sum_{i=1}^{N} \varepsilon_{i,j} \qquad [5.12]$$

The canonical partition function is therefore:

$$Z_C = \sum_j \exp(-\beta E_j) = \sum_j \exp-\beta\left[\sum_{i=1}^{N} \varepsilon_{i,j}\right] \qquad [5.13]$$

However, instead of performing the calculation in expression [5.13] for all states j of the system, we can successively perform this calculation on all individual states j of each molecule i, which gives:

$$Z_C = \prod_{i=1}^{N}\left\{\sum_j \exp\left(-\beta\varepsilon_j\right)\right\} \qquad [5.14]$$

From the bracket which is the molecular partition function (see the note in section 5.4), we can write:

$$Z_C = z^N \tag{5.15}$$

In the case of several independent constituents, A, B, etc., either belonging to different phases or the same phase but without interaction between molecules of two different constituents, relation [5.15] becomes:

$$Z_C = \prod_{\text{A}} Z_{C(\text{A})} = \prod_{\text{A}} z_{\text{A}}^{\text{N}_\text{A}} \tag{5.16}$$

So, for ensembles of discernable molecules, the canonical partition function can be calculated from the molecular partition functions using relations [5.15] or [5.16] depending on whether the system has one or several constituents. These relations will be used for crystals in which the atoms (or molecules) are localized to network nodes.

5.3.2. *Canonical partition functions of indiscernible molecules*

It is only through a complex calculation that we can link the canonical partition functions of indiscernible molecules to their molecular partition functions. Fowler, proceeding by trial and error by *a posteriori* proving the result using the properties of the Helmholtz function F, helped form the following relations:

– in the case of A particles that obey Bose-Einstein statistics:

$$\ln Z_{C(\text{A})} = -\sum_{\text{i}} g_{i(\text{A})} \ln\left[1 - \exp\left(-\alpha_\text{A} - \beta\varepsilon_i\right)\right] + N_\text{A}\alpha_\text{A} \tag{5.17}$$

– in the case of A particles that obey Fermi-Dirac statistics:

$$\ln Z_{C(\text{A})} = \sum_{\text{i}} g_{i(\text{A})} \ln\left[1 + \exp\left(-\alpha_\text{A} - \beta\varepsilon_i\right)\right] + N_\text{A}\alpha_\text{A} \tag{5.18}$$

By examining expressions [5.17] and [5.18], we can write in the general form:

$$\ln Z_{C(\text{A})} = \varepsilon \sum_{i} g_{i(\text{A})} \ln\left[1 + \varepsilon \exp\left(-\alpha_\text{A} - \beta\varepsilon_i\right)\right] + N_\text{A}\alpha_\text{A} \tag{5.19}$$

with convention $\varepsilon = -1$ for Bose-Einstein statistics and $\varepsilon = +1$ for Fermi-Dirac statistics.

We know that the three statistics merge in the "classical limiting case", which will be described by relation [5.19] with $\varepsilon = +1$.

Taking into account the approximations, we can write:

$$\ln\left[1+\varepsilon\exp\left(-\alpha_{\mathrm{A}}-\beta\varepsilon_i\right)\right]\cong\varepsilon\exp\left(-\alpha_{\mathrm{A}}-\beta\varepsilon_i\right) \qquad [5.20]$$

Applying this approximation to relation [5.19] therefore gives:

$$\ln Z_{C(\mathrm{A})}=\varepsilon\sum_i g_{i(\mathrm{A})}\exp\left(-\alpha_{\mathrm{A}}-\beta\varepsilon_i\right)+N_{\mathrm{A}}\alpha_{\mathrm{A}} \qquad [5.21]$$

where: $\varepsilon^2 = 1$.

Taking into account the definition of the molecular partition function (relation [4.49]), this expression becomes:

$$\ln Z_{C(\mathrm{A})}=\exp(-\alpha_{\mathrm{A}})z_{\mathrm{A}}+N_{\mathrm{A}}\alpha_{\mathrm{A}} \qquad [5.22]$$

By introducing equations [4.91] and [4.92], adhering to the first terms of the equations, we obtain:

$$\ln Z_{C(\mathrm{A})}=N_{\mathrm{A}}(1+\alpha_{\mathrm{A}})=N_{\mathrm{A}}\left(1+\ln\frac{z_{\mathrm{A}}}{N_{\mathrm{A}}}\right) \qquad [5.23]$$

Using Stirling's equation [4.1] we obtain:

$$\ln Z_{C(\mathrm{A})}=N_{\mathrm{A}}\ln z_{\mathrm{A}}-\ln\left(N_{\mathrm{A}}!\right) \qquad [5.24]$$

and therefore:

$$Z_{C(\mathrm{A})}=\frac{z_{\mathrm{A}}^{N_{\mathrm{A}}}}{N_{\mathrm{A}}!} \qquad [5.25]$$

For several constituents according to the first equality [5.16] we obtain:

$$Z_C = \prod_{\text{A}} Z_{C(\text{A})} = \prod_{A} \frac{z_{\text{A}}^{N_{\text{A}}}}{N_{\text{A}}!} \qquad [5.26]$$

The product is extended to all constituents.

Relations [5.25] or [5.26] will essentially be used for gases and liquids whose molecules are indiscernible.

5.4. Thermodynamic functions and the canonical partition function

Now we aim to calculate the thermodynamic functions and the thermodynamic coefficients knowing the canonical partition function of the system. We will begin with the internal energy and the entropy, and then once these two functions are known, the other thermodynamic functions or their derivatives can easily be calculated.

5.4.1. *Expression of internal energy*

When establishing the statistics of the canonical ensemble, we have chosen as the energies of a system at the lowest level, therefore the internal energy must take into account the energy at 0 K and therefore the internal energy, which is the energy of the average element in the canonical ensemble, according to relation [4.2], is:

$$U = U(0) + \frac{E_C}{N_C} \qquad [5.27]$$

Using relations [5.2] and [5.6], it becomes:

$$U - U(0) = \frac{E_C}{N_C} = \frac{1}{N_C}\sum_j n_j E_j = \frac{1}{Z_C \sum_j E_j \exp(-\beta E_j)} \qquad [5.28]$$

Note that:

$$\frac{\mathrm{d}\left[\exp\left(-\beta E_j\right)\right]}{\mathrm{d}\beta} = -E_j \exp\left(-\beta E_j\right) \quad [5.29]$$

which using relation [5.28] gives:

$$U - U(0) = -\frac{1}{Z_C}\frac{\mathrm{d}\left[\sum_j E_j \exp(-\beta E_j)\right]}{\mathrm{d}\beta} = -\frac{1}{Z_C}\frac{\mathrm{d}Z_C}{\mathrm{d}\beta} \quad [5.30]$$

In expression [5.30], the partition function Z_C may depend on variables other than temperature, such as volume, for example; the derivative here is therefore a partial derivative with regard to temperature, thus keeping the other variables constant, giving the expression of the internal energy:

$$U - U(0) = -\frac{\partial \ln Z_C}{\partial \beta} = k_B T^2 \frac{\partial \ln Z_C}{\partial T} \quad [5.31]$$

We find that only the temperature and the canonical partition function must be known to calculate the internal energy of the system.

In the case of a mixture with several constituents A, B, C, etc., we easily obtain:

$$U - U(0) = -\sum_{\mathrm{A}} \frac{\partial \ln Z_{C(\mathrm{A})}}{\partial \beta} = \mathrm{k_B} T^2 \sum_{\mathrm{A}} \frac{\partial \ln Z_{C(\mathrm{A})}}{\partial T} \quad [5.32]$$

5.4.2. *Entropy and canonical partition functions*

We are going to express the average entropy of a system using the partition function of the canonical ensemble.

From relation [4.41], we can write:

$$S = \frac{S_C}{N_C} = \frac{\mathrm{k_B}}{N_C} \ln \Omega_C \quad [5.33]$$

By applying relation [4.7] to the canonical ensemble, it becomes:

$$S = -\mathrm{k_B} \sum_j \frac{n_j}{N_C} \ln \frac{n_j}{N_C} \quad [5.34]$$

By incorporating this into equation [5.11], we obtain:

$$S = -\mathrm{k_B} \sum_j p_j \ln p_j \quad [5.35]$$

This equation [5.11] can also be written as follows:

$$\ln p_j = -\beta E_j - \ln Z_C \quad [5.36]$$

i.e. the expression for entropy of the system:

$$S = -\mathrm{k_B} \left[-\beta \sum_j p_j \ln p_j - \ln Z_C \sum_j p_j \right] \quad [5.37]$$

Thus, we know that:

$$\sum_j p_j = 1$$

Now by taking into account relations [5.11] and [5.28], we obtain:

$$\sum_j p_j E_j = U - U(0) \quad [5.38]$$

i.e. the entropy is:

$$S = \mathrm{k_B} \beta [U - U(0)] + \mathrm{k_B} \ln Z_C \quad [5.39]$$

Determining the internal energy using relation [5.31], we obtain a second form in which to express the entropy:

$$S = \mathrm{k_B} \left[\ln Z_C - \frac{1}{T} \frac{\partial \ln Z_C}{\partial \ln \beta} \right] \quad [5.40]$$

So, just like the internal energy, entropy is expressed as a function of the only canonical partition function of the system.

In the case of a mixture with several constituents, we obtain:

$$S = k_B \sum_{A} \left[\ln Z_{C(A)} - \frac{1}{T} \frac{\partial \ln Z_{C(A)}}{\partial \ln \beta} \right] \quad [5.41]$$

Having obtained the internal energy and the entropy, we can conclude that the canonical partition function of the system can completely define the system in the thermodynamic plane. We can therefore express any thermodynamic function using the canonical partition function, expressed as variables amount of matter (N), volume (V) and temperature (T), i.e. the canonical variables associated with the Helmholtz function F.

5.4.3. *Expressing other thermodynamic functions and thermodynamic coefficients in the canonical ensemble*

With the expressions of internal energy and entropy, we can calculate all other thermodynamic functions defined by them. This is the same for the primary and secondary partial derivatives of these thermodynamic functions, i.e. the conjugate variables of the problem variables including pressure (or its opposite), entropy, the chemical potential and the thermodynamic coefficients which are the secondary derivatives of thermodynamic functions. Below are the results, easily obtained for any of the variables.

5.4.3.1. *Helmholtz free energy*

Using relations [5.31] and [5.41], we calculate:

$$F - F(0) = U - U(0) - TS = -\frac{\ln Z_C}{\beta} = -k_B T \ln Z_C \quad [5.42]$$

If it is a mixture of several constituents, A, B, C etc., we obtain:

$$F - F(0) = -\frac{\sum_{A} \ln Z_{C(A)}}{\beta} = -k_B T \sum_{A} \ln Z_{C(A)} \quad [5.43]$$

The equation is extended to all constituents.

5.4.3.2. *Pressure*

Using equation [5.43], we calculate:

$$P = -\left(\frac{\partial F}{\partial V}\right)_{T,N_A,..} = \frac{1}{\beta}\left(\frac{\partial \ln Z_C}{\partial V}\right)_{T,N_A,..} = k_B T\left(\frac{\partial \ln Z_C}{\partial V}\right)_{T,N_A,..} \quad [5.44]$$

With several constituents:

$$P = \frac{1}{\beta}\sum_A \left(\frac{\partial \ln Z_{C(A)}}{\partial V_A}\right)_{T,N_A,..} = k_B T \sum_A \left(\frac{\partial \ln Z_{C(A)}}{\partial V_A}\right)_{T,N_A,..} \quad [5.45]$$

5.4.3.3. *Gibbs free energy*

Using equations [5.43] and [5.45] we can write:

$$G - G(0) = F - F(0) - PV = -\frac{\ln Z_C}{\beta} + \frac{V}{\beta}\left(\frac{\partial \ln Z_C}{\partial V}\right)_{T,N} \quad [5.46]$$

with several constituents:

$$G - G(0) = -\frac{\sum_A \ln Z_{C(A)}}{\beta} + \frac{V}{\beta}\sum_A \left(\frac{\partial \ln Z_{C(A)}}{\partial V_A}\right)_{T,N} \quad [5.47]$$

5.4.3.4. *Chemical potential*

Using equation [5.43], we calculate:

$$\mu_A = \left(\frac{\partial F}{\partial N_A}\right)_{T,V,N_B..} = -\frac{1}{\beta}\left(\frac{\partial \ln Z_{C(A)}}{\partial N_A}\right)_{T,V,N_B..} \quad [5.48]$$

5.4.3.5. *Heat capacity at constant volume*

We are going to detail the calculation method of the thermodynamic coefficient by using the example calculation for the specific heat capacity at constant volume. For this, we can use either the secondary derivative of the characteristic function F, or, more simply, calculate the specific heat capacity at volume constant from the internal energy (equation [5.31]):

$$C_{V_A} = \left(\frac{\partial U_A}{\partial T}\right)_{N,V} = 2k_B T \frac{\partial \ln Z}{\partial T} + k_B T^2 \frac{\partial^2 \ln Z}{\partial T^2} \qquad [5.49]$$

Ultimately, we know that in thermodynamics, the characteristic function in the variables' temperature, volume, quantities of matter, etc., is the Helmholtz function F; therefore, knowing F provides a full description of the system. As shown by relation [5.43], if only the canonical partition function is known, the whole system can be characterized. The calculation of such functions is therefore essential.

NOTE (molar variables).– In the expressions of molecular and canonical partition functions that we gave for a certain number of molecules, the number of particles is denoted by N_A for constituent A.

To obtain the molar values of thermodynamic functions, it is worth choosing Avogadro's number (N_a) for N_A.

To obtain the value of a function for an amount of n (in moles) of constituent, we use the product nN_a for N_A.

In these two cases, the expression of β will always be 1/RT.

5.5. Absolute activity of a constituent

The canonical partition function relative to constituent A is *a priori* a function of β (or T), of volume and amount N_A of the constituent. We can therefore write the differential of this partition function in the form:

$$\begin{aligned} d \ln Z_{C(A)} &= \frac{\partial \ln Z_{C(A)}}{\partial \beta} d\beta + \sum_i \frac{\partial \ln Z_{C(A)}}{\partial \varepsilon_i} \frac{\partial \varepsilon_i}{\partial V} dV \\ &+ \frac{\partial \ln Z_{C(A)}}{\partial N_A} dN_A + \frac{\partial \ln Z_{C(A)}}{\partial \alpha_A} \left[\frac{\partial \alpha_A}{\partial \beta} d\beta + \frac{\partial \alpha_A}{\partial V} dV + \frac{\partial \alpha_A}{\partial N_A} dN_A \right] \end{aligned} \qquad [5.50]$$

By differentiating expression [5.19] we obtain:

$$\frac{\partial \ln Z_{C(A)}}{\partial \alpha_A} = -\varepsilon \sum_i g_i \frac{\varepsilon \exp(-\alpha_A - \beta \varepsilon_i)}{1 + \varepsilon \exp(-\alpha_A - \beta \varepsilon_i)} + N_A \qquad [5.51]$$

Applying statistics gives:

$$N_A = \sum_i n_i = \varepsilon \sum_i g_i \frac{\varepsilon \exp(-\alpha_A - \beta \varepsilon_i)}{1 + \varepsilon \exp(-\alpha_A - \beta \varepsilon_i)} \qquad [5.52]$$

Using equality [5.50] we obtain:

$$\frac{\partial \ln Z_{C(A)}}{\partial \alpha_A} = N_A - N_A = 0 \qquad [5.53]$$

Everything happens as if, in the first term of equation [5.19], term α_A is a constant and therefore by differentiating this expression with this condition, it becomes:

$$\frac{\partial \ln Z_{C(A)}}{\partial N_A} = \alpha_A \qquad [5.54]$$

Remembering that the chemical potential is given by relation [5.48], we find that we can write:

$$\mu_A = -\frac{\alpha_A}{\beta} \qquad [5.55]$$

This relation is true regardless of the statistic used, whether Maxwell–Boltzmann, Bose–Einstein or Fermi–Dirac statistics.

NOTE.– The chemical potential here is the generalized chemical potential which, for example, for the electrically charged particles, is the electrochemical potential $\overline{\mu}_A$.

In certain developments, we sometimes introduce the term λ_A defined by:

$$\ln \lambda_A = -\alpha_A \qquad [5.56]$$

The chemical potential therefore takes the form:

$$\mu_A = k_B T \ln \lambda_A \qquad [5.57]$$

i.e.:

$$\lambda_A = \exp\left(\frac{\mu_A}{k_B T}\right) \qquad [5.58]$$

The variable λ_A is called *absolute activity* of constituent A (not to be confused with the activity).

5.6. Other ensembles of systems and associated characteristic functions

In this chapter we have described a canonical ensemble. In fact, an ensemble is chosen according to the set of independent variables selected, i.e. volume, temperature and amount of matter of the canonical ensemble. For this ensemble, we defined a canonical partition function by relation [5.8].

Moreover, as seen in Chapter 1, each set of variables has a corresponding characteristic function, which is the thermodynamic potential. We saw that this is the Helmholtz free energy for variables linked to a canonical ensemble. Also note that this function is linked to the statistical description from the relation [5.42] and to the logarithm of the partition function of the corresponding ensemble.

In the same way as we chose in Chapters 1 and 2 the characteristic function and a characteristic matrix for each set of independent variables E_p, we can link to this set an ensemble, as in section 5.2, such that the relation between the characteristic function and the partition function linked to this ensemble, i.e. in the same form as in relation [5.42].

So, for example, if we choose the set of independent variables: temperature, volume and chemical potential, we show that the characteristic function is the product $-PV$; the corresponding ensemble is called the *grand canonical ensemble*. The function $-PV$ will be defined in the statistical plane by relation:

$$-PV = \frac{1}{\beta} \ln Z_{\{T,-P,\mu\}} \qquad [5.59]$$

$Z_{\{T,-P,\mu\}}$ is the grand canonical partition function. It is worth showing the equivalence of the statistical definition of the product PV given by relation [5.59] and the macroscopic definition given by partial derivatives. It is then easy to calculate the expression of any thermodynamic function and any thermodynamic coefficient from the partition function linked to the new ensemble. Finally, it is worth linking the partition function of the new ensemble to molecular partition functions.

This ensemble is mainly used when choosing the chemical potential as a variable is interesting such as, for example, in the case of equilibria between phases. The chemical potentials are therefore constant between each phase. For the description of a phase, the canonical ensemble IS preferred.

We will no longer develop the grand canonical ensemble as it is primarily used to study phase transformations directly from statistical thermodynamics, which we are not going to do. In fact, phases can be sufficiently described and potential functions can be expressed using the molecule's characteristics to study the equilibria between phases using potential functions. The direct transition from microscopic descriptions to the equilibria of phases requires a large mathematical tool, which provides no additional information about the modeling of these transformations.

We can also choose, as variables, the ensemble:

$$\mathcal{E}_p = \{T,-P,N\}$$

The corresponding characteristic function will be the Gibbs function defined in the statistical plane by:

$$G = \frac{1}{\beta} \ln Z_{\{T,-P,N\}} \qquad [5.60]$$

$Z_{\{T,-P,N\}}$ will be the partition function linked to the new ensemble.

NOTE.– The isolated ensemble, in which the number of molecules, volume and energy are chosen as variables and kept constant, is the micro-canonical that we came across in Chapter 4. Its potential function is entropy. The corresponding partition function is simply the molecular partition function. However, calculating thermodynamic functions is tricky, hence, the preference to use the canonical ensemble when attempting to describe phases.

6

Molecular Partition Functions

6.1. Definition of the molecular partition function

The molecular partition function is defined, for an ensemble composed of many particles of molecular dimensions each with the energy ε_i, by the equation:

$$z = \sum_i g_i \exp\left(-\frac{\varepsilon_i}{\mathrm{k_B}T}\right) = \sum_i g_i \exp(-\varepsilon_i \beta) \qquad [6.1]$$

6.2. Decomposition of the molecular partition function into partial partition functions

The molecular partition function for a system includes terms that relate to different forms of energy: nuclear, electronic, vibrational energy of molecules, their rotational energy, their translational energy and interaction energies between different molecules.

To simplify this, we know that different forms of energy, for a molecule, are independent (however, there are exceptions, e.g. the interdependence of vibrational and rotational energies of a molecule). Under these conditions, we can write the total energy of a molecule as the sum of different contributions by different forms of energy: nuclear ε_n, electronic ε_e, vibrational ε_v, rotational ε_r, translational ε_t and interaction ε_I, i.e.:

$$\varepsilon = \varepsilon_n + \varepsilon_e + \varepsilon_v + \varepsilon_r + \varepsilon_t + \varepsilon_I \qquad [6.2]$$

The partition function becomes:

$$z = \sum_{i_n} \exp\left(-\frac{\varepsilon_{i_n}}{k_B T}\right) \sum_{i_e} \exp\left(-\frac{\varepsilon_{i_e}}{k_B T}\right) \sum_{i_v} \exp\left(-\frac{\varepsilon_{i_v}}{k_B T}\right)$$
$$\sum_{i_r} \exp\left(-\frac{\varepsilon_{i_r}}{k_B T}\right) \sum_{i_t} \exp\left(-\frac{\varepsilon_{i_t}}{k_B T}\right) \sum_{i_I} \exp\left(-\frac{\varepsilon_{i_I}}{k_B T}\right) \quad [6.3]$$

This shows the partial partition functions related to the different forms of energy:

$$z_n = \sum_{i_n} \exp\left(-\frac{\varepsilon_{i_n}}{k_B T}\right), z_e = \sum_{i_e} \exp\left(-\frac{\varepsilon_{i_e}}{k_B T}\right), z_v = \sum_{i_v} \exp\left(-\frac{\varepsilon_{i_v}}{k_B T}\right)$$

$$z_r = \sum_{i_r} \exp\left(-\frac{\varepsilon_{i_r}}{k_B T}\right), z_t = \sum_{i_t} \exp\left(-\frac{\varepsilon_{i_t}}{k_B T}\right), z_I = \sum_{i_I} \exp\left(-\frac{\varepsilon_{i_I}}{k_B T}\right)$$

[6.4]

The overall partition function takes the form of the product of the partial partition functions:

$$z = z_n z_e z_v z_r z_t z_I \quad [6.5]$$

Calculating partial partition functions requires that the number of degrees of freedom for each type of molecule movement is known.

If a molecule is composed of N atoms, it has a total of $3N$ degrees of freedom. In the case of a liquid or gas, molecules usually have three degrees of translational freedom. A non-linear molecule has three degrees of rotational freedom whereas a linear molecule has only two. The number of degrees of vibrational freedom is easily deduced:

– linear molecules: $3N$-5 degrees of vibrational freedom;

– non-linear molecules: $3N$-6 degrees of vibrational freedom.

For example, in a gas or a liquid, a diatomic molecule ($N = 2$) that is obligatorily linear will have six degrees of freedom – three translational, two rotational and one degree of vibrational freedom.

In the case of a crystallized solid, there are only vibrational degrees of freedom, i.e. $3N$, as the molecules are fixed on the sites of the crystal lattice.

To calculate the energies, we must also choose a level of origin; it is worth choosing the first level corresponding to each movement:

– for the kinetic energy, we choose zero velocity as the origin;

– for the potential interaction energy between particles, we often choose infinitely remote particles as the origin;

– for the rotational energy, we choose the level corresponding to the zero value of the quantum rotational number: $j = 0$;

– for the vibration energy, we choose either the minimum of the potential energy curve or the residual level corresponding to a zero value of the vibrational quantum number (v = 0) i.e. $\varepsilon_0 = 1/2h\nu$ (ν is the fundamental vibrational frequency).

6.3. Energy level and thermal agitation

Normally, calculating partition functions requires the use of energy levels compatible with, on the one hand, results of quantum mechanics, and on the other hand, summations of relation [6.3]. However, the calculation can be simplified in two extreme cases.

Let us denote ε_0 and ε_1 as the two first energy levels that correspond to one degree of freedom. If the difference between these two levels is very large with regard to thermal agitation, i.e. the product $\mathrm{k_B}T$, therefore if:

$$\varepsilon_1 - \varepsilon_0 >> \mathrm{k_B}T \qquad [6.6]$$

The partition function corresponding to this degree of freedom is reduced to the first term of equation [6.1], i.e.:

$$z = g_0 \exp(-\varepsilon_0 \beta) \qquad [6.7]$$

If we choose the first level, i.e. $\varepsilon_0 = 0$ as the origin of the energies of this degree of freedom, then the partition function simply becomes:

$$z = g_0 \qquad [6.8]$$

This degree of freedom is said to be non-excited. This is found to be so for electronic and nuclear degrees of freedom.

Now, if the difference between the two first energy levels is very small with regard to the thermal agitation, it means that:

$$\varepsilon_1 - \varepsilon_0 << \mathrm{k_B}T \qquad [6.9]$$

Also, if the approximation conditions of classical statistics are satisfied (see relation [4.73]), we say that we are dealing with a classical degree of freedom. This case is most frequently found with translational, rotational and vibrational energies.

In the cases above, the different energy levels are very narrow, and rather than tabulating them one by one, it is better to group them in a single state; for one degree of freedom, all levels included in the element with area $\mathrm{d}p_x.\mathrm{d}x$, composed of the product of the linear momentum and a positional element in the space of phases. However, as we are concerned with quantum statistics, according to the Heisenberg principle, there is only one state per quantum box with area h (Planck's constant) in the same axis system; we will therefore attribute to this single state, a statistical weight defined by:

$$g_i = \frac{\mathrm{d}\,p_x.\mathrm{d}\,x}{\mathrm{h}} \qquad [6.10]$$

As there are many very narrow energy levels, the equation that appears in the definition of the partition function can be replaced by an integral and for one degree of freedom we obtain the partition function given by:

$$z = \int_{-\infty}^{\infty}\int_{-\infty}^{\infty} \exp\left(-\frac{\varepsilon}{\mathrm{k_B}T}\right)\frac{\mathrm{d}\,p_x.\mathrm{d}\,x}{\mathrm{h}} \qquad [6.11]$$

This is often found for translational, rotational and vibrational energies.

6.4. Translational partition functions

To the three directions x_1, x_2, and x_3 of space and for the translation particles, we will apply relation [6.10]; the statistical weight will therefore be:

$$g_i = \frac{\mathrm{d}\,p_{x_1}.\mathrm{d}\,p_{x_2}.\mathrm{d}\,p_{x_3}.\mathrm{d}\,x_1.\mathrm{d}\,x_2.\mathrm{d}\,x_3}{\mathrm{h}^3} \qquad [6.12]$$

Now let us apply relation [6.11] to the three directions and the three amounts of movement:

$$z_t = \frac{1}{\mathrm{h}^3}\int_{-\infty}^{\infty}\int_{-\infty}^{\infty}\int_{-\infty}^{\infty}\int_{-\infty}^{\infty}\int_{-\infty}^{\infty}\int_{-\infty}^{\infty}\exp\left(-\frac{\varepsilon}{\mathrm{k_B}T}\right)\mathrm{d}\,p_{x_1}.\mathrm{d}\,p_{x_2}.\mathrm{d}\,p_{x_3}.\mathrm{d}\,x_1.\mathrm{d}\,x_2.\mathrm{d}\,x_3$$

[6.13]

We are now are going to assume, as we have done previously, that the degrees of freedom are independent and assess the energy involved. For this we will consider two cases where the particle is or is not subject to a potential energy. In these cases, the particle is forced to remain in a set volume V.

6.4.1. *Translational partition function with the only constraint being the recipient*

The particle is not subject to any constraint other than only being able to move in a given volume. The two energetic terms must be distinguished:

– the kinetic energy explained using components of the linear momentum is written:

$$\varepsilon_c = \frac{p_{x_1}^2 + p_{x_2}^2 + p_{x_3}^2}{2m} \qquad [6.14]$$

– the potential energy that constrains a particle to a free volume. Let v_f be the available volume for one particle, the total volume available for molecules of the system is Nv_f and for the potential energy we can write the conditions:

$$\varepsilon_p(x_1, x_2, x_3) = \begin{cases} \infty \ \textbf{if}\ x_1, x_2 \ \textit{and}\ x_3 \ \textbf{outside}\ Nv_f \\ 0 \ \textbf{if}\ x_1, x_2 \ \textit{and}\ x_3 \ \textbf{inside}\ Nv_f \end{cases} \qquad [6.15]$$

By separating the degrees of freedom, relation [6.13] becomes:

$$z_t = \frac{1}{h^3} \int_{-\infty}^{\infty} \exp\left(-\frac{p_{x_1}^2}{2m\mathrm{k_B}T}\right) \mathrm{d}p_{x_1} \int_{-\infty}^{\infty} \exp\left(-\frac{p_{x_2}^2}{2m\mathrm{k_B}T}\right) \mathrm{d}p_{x_2}$$
$$\int_{-\infty}^{\infty} \exp\left(-\frac{p_{x_3}^2}{2m\mathrm{k_B}T}\right) \mathrm{d}p_{x_3} \iiint_{Nv_f} \mathrm{d}x_1 \mathrm{d}x_2 \mathrm{d}x_3 \qquad [6.16]$$

Taking into account the definition of the error function, we know the integral:

$$\int_{-\infty}^{\infty} \exp\left(-u^2\right) \mathrm{d}u = \sqrt{\pi} \qquad [6.17]$$

By changing the variable for each direction i of space in the three first integrals of [6.16]:

$$u_i = \frac{p_i}{2m\mathrm{k_B}T} \qquad [6.18]$$

Expression of the translational partition function is calculated by:

$$z_t = Nv_f \left(\frac{2\pi m\mathrm{k_B}T}{\mathrm{h}^2}\right)^{3/2} \qquad [6.19]$$

This function is often written in the form:

$$z_t = \frac{Nv_f}{\Lambda^3} \qquad [6.20]$$

In which function Λ, which has the dimensions of a length, is defined by:

$$\Lambda = \frac{\mathrm{h}}{\left(2\pi m\mathrm{k_B}T\right)^{1/2}} \qquad [6.21]$$

This length is called the thermal wavelength of the molecule.

If we consider the case of an ideal gas, the molecules are not subject to any force and they are treated like dimensionless points; the free volume for one molecule is equal to the volume of the container divided by the number of molecules, and therefore:

$$Nv_f = V \qquad [6.22]$$

According to [6.19], the translational partition function of the ideal gas is therefore:

$$z_{t(pf)} = V\left(\frac{2\pi m k_B T}{h^2}\right)^{3/2} = \frac{V}{\Lambda^3} \qquad [6.23]$$

NOTE.– We show that the condition that allows state equation [6.1] to be replaced by an integral is equivalent to:

$$\Lambda^3 << V \qquad [6.24]$$

which means that the particle should not interact, simultaneously and significantly with the two opposite sides of the container.

6.4.2. *Translational partition function with the constraint being a potential centered and the container walls*

Now let us assume that the forces exerted between the particles are functions of the relative coordinates of these particles. The origin of these energies due to these forces is chosen for infinitely remote particles.

Due to the involvement of all the particles in the system, it is no longer possible to calculate the molecular contribution of the partition function. It is necessary to consider about the ensemble of N particles as a whole and therefore directly determine the canonical partition function.

Let ω_i be the ensemble of the coordinates of positon $(x_1, x_2, x_3)_i$ of particle i and p_i be the ensemble of components of linear momentum (p_{x_1}, p_{x_2}, p_{x_3})$_i$ of the same particle i. The total translational energy E of the system is the sum of the kinetic energy and the potential energy, which is the energy of interaction between particles and between the particles and the container walls:

$$E(\omega_1,\omega_2,..\omega_i,...,\omega_N,p_1,p_2,...,p_i,..,p_N)$$
$$=\frac{p_1^2}{2m}+\frac{p_2^2}{2m}+..\frac{p_N^2}{2m}+E_I(\omega_1,\omega_2,..\omega_i,...,\omega_N) \qquad [6.25]$$

This energy is a function of 6N variables of the space of phases: the coordinates of position ω_i and linear momenta p_i.

Between two particles i and j, there is an interaction energy $\varepsilon_{i,j}$ and assuming that the interaction energy contains only terms such as $\varepsilon_{i,j}$, even if several molecules are mutually interacting. This assumption is known as the *paired interaction model.*

If we count the ensemble of paired particles capable of interacting, this is the number of combinations with repetition of N particles 2 by 2, i.e. $N(N-1)/2$.

For indistinguishable particles, the translational canonical partition function can be written in the classical approximation, taking into account relations [5.25] and [6.13]:

$$Z_{C(t)}=\frac{1}{N!}\left(\frac{1}{\mathrm{h}^3}\right)^N\int\limits_{Nv_f}\int...\int\exp-\left[\frac{\sum\limits_{i=1}^{N}p_i^2}{2m\mathrm{k_B}T}\prod_{i=1}^{N}\mathrm{d}p_i\right] \qquad [6.26]$$
$$\int\limits_{Nv_f}\int...\int\prod_{i<j}\exp-\frac{\varepsilon_{i,j}}{\mathrm{k_B}T}(\mathrm{d}\omega)^N$$

The integrals are extended to the volume available for the molecules i.e.: Nv_f and the second product contains $N(N-1)/2$ factors.

The calculation for the first integral on the 3N linear moments is identical to that for the translation of an ideal gas whose molecular function z_{tpf} is given by relation [6.23], the translational canonical function can therefore be written as:

$$Z_{C(t)} = \frac{1}{N!}\left(\frac{z_{tpf}}{V}\right)^N \int_{Nv_f}\int ... \int \prod_{i<j} \exp -\frac{\varepsilon_{i,j}}{k_B T}(d\,\omega)^N \quad [6.27]$$

The specific contribution made by the interactions to the canonical partition function can therefore be deduced:

$$I_I = \int_{Nv_f}\int ... \int \prod_{i<j} \exp -\frac{\varepsilon_{i,j}}{k_B T}(d\,\omega)^N \quad [6.28]$$

This term is sometimes called the *configuration integral* because, as we can see, by comparing relations [6.23] and [6.27], this term has the dimensions of a volume.

6.5. Maxwell distribution laws

We are going to apply the law of average distribution of particles (relation [4.2]) to the translational energy in the ideal gas hypothesis, that is to say for molecules considered as dimensionless points we can write, using relations [6.13] and [6.23]:

$$\langle dn \rangle = \frac{N}{V}\left(\frac{2\pi m k_B T}{h^2}\right)^{3/2} \exp\left(-\frac{p_{x_1}^2 + p_{x_2}^2 + p_{x_3}^2}{2mk_B T}\right) \quad [6.29]$$
$$d\,p_{x_1}.d\,p_{x_2}.d\,p_{x_3}.d\,x_1.d\,x_2.d\,x_3$$

This result is one of the forms of the Maxwell distribution law. It can be used for two forms of energy: the potential energy will give the average number of molecules in the volume dV and the kinetic energy will lead to the distribution of molecules according to their velocity.

6.5.1. *Distribution of ideal gas molecules in volume*

Applying relation [6.29] to the potential energy gives:

$$\frac{\langle d\,n \rangle}{d\,V} = \frac{N}{V}\left(\frac{2\pi m k_B T}{h^2}\right)^{3/2} \quad [6.30]$$

This expression shows that the distribution of molecules of an ideal gas is uniform in the volume in the absence of external forces. We can deduce in the same way, by applying the gravitational force, the distribution of molecules with regard to height, i.e. the law of decantation.

6.5.2. *Distribution of ideal gas molecules in velocity*

Now if we apply the Maxwell law to the kinetic energy, we obtain:

$$\langle \mathrm{d}n \rangle = N\left(\frac{m}{2\pi \mathrm{k_B} T}\right)^{3/2} \exp\left(-\frac{p_{x_1}^2 + p_{x_2}^2 + p_{x_3}^2}{2m\mathrm{k_B}T}\right) .\mathrm{d}p_{x_1}.\mathrm{d}p_{x_2}.\mathrm{d}p_{x_3} \quad [6.31]$$

This expression can be written as the average number of particles with velocity components between v_{x_1} and $v_{x_1} + \mathrm{d}v_{x_1}$ and v_{x_2} and $v_{x_2} + \mathrm{d}v_{x_2}$ and v_{x_3} and $v_{x_3} + \mathrm{d}v_{x_3}$:

$$\langle dn \rangle = N\left(\frac{m}{2\pi \mathrm{k_B} T}\right)^{3/2} \exp\left(-\frac{m\left(v_{x_1}^2 + v_{x_2}^2 + v_{x_3}^2\right)}{2\mathrm{k_B}T}\right) .\mathrm{d}v_{x_1}.\mathrm{d}v_{x_2}.\mathrm{d}v_{x_3} \quad [6.32]$$

It is often worth writing this Maxwell law relative to the velocity components in the form of a distribution function $g(v_{x_1}v_{x_2}, v_{x_3})$ such that the product $g(v_{x_1}v_{x_2}, v_x)\mathrm{d}v_x\, \mathrm{d}v_{x_2}\, \mathrm{d}v_{x_3}$ represents the fraction of all the particles whose velocity components are between the intervals v_{x_1} and $v_{x_1} + \mathrm{d}v_{x_1}$, between v_{x_2} and $v_{x_2} + \mathrm{d}v_{x_2}$ and between v_{x_3} and $v_{x_3} + \mathrm{d}v_{x_3}$, we can therefore write according to [6.32]:

$$g(v_x v_y, v_z) = \left(\frac{m}{2\pi \mathrm{k_B} T}\right)^{3/2} \exp\left(-\frac{m\left(v_x^2 + v_{x_2}^2 + v_{x_3}^2\right)}{2\mathrm{k_B}T}\right) \quad [6.33]$$

It is often useful to determine the velocity in a single direction. The number of particles whose component according to Ox_1 is between v_{x_1} and $v_{x_1} + \mathrm{d}\, v_{x_1}$, is given by:

$$\left\langle \mathrm{d}\, n_{v_x} \right\rangle = N \left(\frac{m}{2\pi \mathrm{k_B} T} \right)^{1/2} \exp\left(-\frac{m v_x^2}{2\mathrm{k_B} T} \right) \mathrm{d}\, v_{x_2} \qquad [6.34]$$

The corresponding distribution function $g(v_x)$ will therefore be:

$$g(v_x) = \left(\frac{m}{2\pi \mathrm{k_B} T} \right)^{1/2} \exp\left(-\frac{m v_{x_1}^2}{2\mathrm{k_B} T} \right) \qquad [6.35]$$

Finally, it is sometimes useful to know the distribution of the particles according to their overall velocity v. For this we move onto spherical coordinates ρ, θ, ϕ and we are aiming to find out the average number of particles whose new velocity components are between the intervals v_ρ and $v_\rho + \mathrm{d}v_\rho$, v_θ and $v_\theta + \mathrm{d}v_\theta$, and v_ϕ and $v_\phi + \mathrm{d}v_\phi$. We obtain by changing the referential:

$$\left\langle \mathrm{d}n_{p,\theta,\varphi} \right\rangle = N \left(\frac{m}{2\pi \mathrm{k_B} T} \right)^{3/2} \exp\left(-\frac{v_\rho^2}{2m\mathrm{k_B} T} \right) v^2 .\sin\theta \,\mathrm{d}\rho.\mathrm{d}\theta.\mathrm{d}\varphi \qquad [6.36]$$

By integrating this relation between 0 and π for θ and 0 and 2π for ϕ, we obtain the number of particles whose velocity component is between v_ρ and $v_\rho + \mathrm{d}v_\rho$, which gives:

$$\left\langle \mathrm{d}\, n_v \right\rangle = 4\pi N \left(\frac{m}{2\pi \mathrm{k_B} T} \right)^{3/2} \exp\left(\frac{v_\rho^2}{2m\mathrm{k_B} T} \right) v_\rho^{\ 2}\, \mathrm{d}\, v_\rho \qquad [6.37]$$

This expression is primarily used in kinetic gas theory to calculate the pressure of the ideal gas per number of collisions with the walls of the container; in collision theory in chemical kinetics to calculate the frequency of collisions between two molecules of an ensemble and in the calculation of the adsorption velocity of a gas onto a solid from the number of collisions to the surface of the solid.

6.6. Internal partition functions

Internal partition functions are relative to the vibrational, rotational, electronic and nuclear movements. We usually find that these movements are independent, although we know that this is not always so between the vibrations and rotations. We also find that the forces exerted on the molecule by the exterior have no influence on these internal degrees of freedom.

6.6.1. *Vibrational partition function*

We are going to distinguish between diatomic molecules and molecules with more than two atoms.

6.6.1.1. *Diatomic molecules*

In the case of a diatomic molecule, there is only one fundamental vibration to consider. Assuming the vibration to be harmonic (harmonic oscillator) and independent of rotation, the vibrational energy is given by:

$$\varepsilon_v = \left(\mathrm{v} + \frac{1}{2}\right) \mathrm{h}\nu \qquad [6.38]$$

In this relation, v is the quantum vibrational number and takes the values 0, 1, 2, 3, etc. and ν is the vibrational frequency.

Each vibration has a unit statistic weight; the vibrational partition function will be:

$$z_v = \sum \exp\left[-\left(\mathrm{v} + \frac{1}{2}\right)\frac{\mathrm{h}\nu}{\mathrm{k_B}T}\right] \qquad [6.39]$$

We define the characteristic vibrational temperature, called the *Einstein temperature* by:

$$\Theta_{\mathrm{E}} = \frac{\mathrm{h}\nu}{\mathrm{k_B}} \qquad [6.40]$$

The vibrational partition function becomes:

$$z_v = \sum \exp\left[-\left(v + \frac{1}{2}\right)\frac{\Theta_E}{T}\right] \qquad [6.41]$$

To assess the sum, let:

$$\exp\left[-\frac{\Theta_E}{T}\right] = x$$

The sum is then written:

$$\sum \exp\left[-(v)\frac{\Theta_E}{T}\right] = x(0 + 1 + 2x + 3x^2 + \ldots + px^{p-1} + \ldots$$

Note that:

$$\frac{d}{dx}(1 + 2x + 3x^2 + \ldots + px^{p-1} + \ldots) = \frac{d}{dx}\left(\frac{1}{1-x}\right) = \frac{1}{(1-x)^2}$$

The sum is therefore:

$$\sum \exp\left[-(v)\frac{\Theta_E}{T}\right] = \frac{1}{\left\{1 + \exp\left[-\frac{\Theta_E}{T}\right]\right\}^2}$$

And the vibrational partition function becomes:

$$z_v = \frac{\exp\left[-\frac{\Theta_E}{2T}\right]}{1 - \exp\left[-\frac{\Theta_E}{T}\right]} \qquad [6.42]$$

In the case where:

$$\frac{\Theta_E}{T} >> 1$$

the vibrational function is simplified to:

$$z_v \cong \frac{T}{\Theta_E} \quad [6.43]$$

	I_2	F_2	HCl	H_2
$\nu(cm^{-1})$	215	892	2990	4400
Θ_E (K)	309	1280	4300	6330

Table 6.1. *Einstein temperature of diatomic gases*

Table 6.1 gives the values of the Einstein temperature for several diatomic gases.

6.6.1.2. *Complex molecules*

Now we must consider several vibrations and assume that different vibrations do not have any influence, which is clearly an approximation. Let ν_1, ν_2, etc. be the vibrational frequencies and Θ_1, Θ_2, etc. the corresponding characteristic temperatures, calculated using relation [6.40]; the vibrational partition function will therefore be in the following form:

$$z_v \cong \prod_E \frac{\exp\left(-\frac{\Theta_E}{2T}\right)}{1-\exp\left(-\frac{\Theta_E}{T}\right)} \quad [6.44]$$

6.6.2. *Rotational partition function*

There is no rotational term to consider for monoatomic molecules since the rotation about an axis passing through the nucleus, where practically the whole mass is concentrated, does not involve any energy. We will now consider three cases of polyatomic molecules.

6.6.2.1. *Heteronuclear diatomic molecules*

These are diatomic molecules with two different atoms HCl.

If the rotation about an axis passing through two nuclei does not involve any energy, there is therefore room to consider that the rotations occur about an axis perpendicular to this line.

According to Schrödinger's equation, the energy of the rigid rotor of the moment of inertia I (see Appendix A.2) is given by:

$$\varepsilon_r = j(j+1)\frac{\mathrm{h}^2}{8\pi^2 I} \qquad [6.45]$$

j is the rotational quantum number and can only take the values 0, 1, 2, etc.

Each energy level has a multiplicity g given by:

$$g = 2j + 1 \qquad [6.46]$$

The rotational partition function is given by:

$$z_r = \sum_j (2j+1)\exp\left[-j(j+1)\frac{\mathrm{h}^2}{8\pi^2 I \mathrm{k_B} T}\right] \qquad [6.47]$$

We call the characteristic rotational temperature, the quantity Θ_r defined by:

$$\Theta_r = \frac{\mathrm{h}^2}{8\pi^2 I \mathrm{k_B}} \qquad [6.48]$$

The rotational function takes the form:

$$z_r = \sum_j (2j+1)\exp\left[-j(j+1)\frac{\Theta_r}{T}\right] \qquad [6.49]$$

If the ratio Θ_r/T is considered to be quite small, the molecules are therefore distributed over a large number of rotational states and the equation can be replaced by an integral, i.e.:

$$z_{rot} = \int_0^{\infty} (2j+1)\exp\left[-j(j+1)\frac{\Theta_r}{T}\right]\mathrm{d}j \qquad [6.50]$$

By the integration of $j(j+1)\dfrac{\Theta_r}{T} = x$:

$$z_r = \frac{T}{\Theta_r} = \frac{8\pi^2 I k_B T}{h^2} \quad [6.51]$$

6.6.2.2. *Homonuclear diatomic molecules*

If both atoms are identical, the total rotation is no longer 360° but 180° since the molecule superposes itself and therefore if we take the previous calculation, we will have therefore counted twice as many states; the rotational partition function is therefore:

$$z_r = \frac{T}{2\Theta_r} = \frac{4\pi^2 I k_B T}{h^2} \quad [6.52]$$

6.6.2.3. *Complex molecules*

In the case of molecules with more than two atoms, if the molecule is linear, the previous results subsist; if the molecule is not linear, we must consider the three moments of inertia I_1, I_2 and I_3 with regard to the three orthogonal axes passing through the center of gravity. In nearly all cases, we can equate sums to integrals and for the rotational partition function we obtain:

$$z_r = \frac{8\pi^2 (2\pi k_B T)^{3/2} \sqrt{I_1 I_2 I_3}}{\sigma h^2} \quad [6.53]$$

The term σ is a term of symmetry whose origin will shall see in the following section (see section 6.6.3).

6.6.3. *Nuclear partition function and correction of symmetry due to nuclear spin*

In the calculation of the nuclear partition function, we must take into account the conditions of symmetry or anti-symmetry of the wave function with regard to the nuclei when they are identical.

If the two nuclei are different, the nuclei a and b are in their energetic state with the lowest statistical weight g_a and g_b so that there are $g_a g_b$ possible nuclear states of molecule AB, therefore:

$$z_n = g_a g_b \qquad [6.54]$$

and if we combine the rotational and nuclear states, we obtain the rotational and nuclear partition function $z_{r,n}$:

$$z_{r,n} = \frac{8\pi^2 I k_B T}{h^2} g_a g_b \qquad [6.55]$$

There is no change compared with what we saw in sections 6.6.1 and 6.6.2.1.

If the two atoms of the molecule are identical, each nucleus can be found in g states; there are therefore g^2 nuclear states of the molecule to take into account.

Moreover, we show that the rotational eigenfunctions are symmetrical with regard to the nuclei if the rotational quantum number j is zero or even anti-symmetrical if it is odd.

This is the same for vibrational eigenfunctions with respect to the vibrational quantum number v.

Firstly, we are going to assume that the molecules are in their lowest vibrational state, therefore v = 0, and the wave functions are symmetrical with regard to the nucleus, which happens at a low enough temperature.

If the nucleus has an odd mass number, the overall wave function is anti-symmetrical with regard to the nuclei; it is the product of all the translational, rotational, vibrational, electronic and nuclear wave functions. With all the translational, vibrational and electronic wave functions being symmetrical, we only have to consider the rotational and nuclear functions, one of which must be symmetrical and the other anti-symmetrical or vice versa. The $g(g-1)/2$ wave functions with anti-symmetrical nuclear spin must have corresponding symmetrical rotational wave functions, i.e. with even values of j the $g(g+1)/2$ wave functions with symmetrical nuclear spin must have corresponding rotational wave functions, i.e. with odd values of j. The combined nuclear-rotational partition function will therefore be:

$$z_{r,n} = \frac{1}{2}g(g-1) \sum_{j=0,2,4,..} (2j+1)\exp\left[-j(j+1)\frac{\Theta_r}{T}\right] + \frac{1}{2}g(g+1) \sum_{j=1,3,5,..} (2j+1)\exp\left[-j(j+1)\frac{\Theta_r}{T}\right] \quad [6.56]$$

If the nucleus has an even mass number, the result is the opposite, i.e.:

$$z_{r,n} = \frac{1}{2}g(g+1) \sum_{j=0,2,4,..} (2j+1)\exp\left[-j(j+1)\frac{\Theta_r}{T}\right] + \frac{1}{2}g(g-1) \sum_{j=1,3,5,..} (2j+1)\exp\left[-j(j+1)\frac{\Theta_r}{T}\right] \quad [6.57]$$

At the ordinary temperature, the ratio Θ_r/T is considered to be small for all molecules and we can therefore say that the two previous equations are almost equal since j is higher and therefore we can write the following for each:

$$z_{r,n} = \frac{1}{2} \sum_{j=0,1,2,..} (2j+1)\exp\left[-j(j+1)\frac{\Theta_r}{T}\right] \quad [6.58]$$

which gives for the rotational nuclear partition function:

$$z_{r,n} = \frac{g^2}{2}\frac{8\pi^2 I k_B T}{h^2} \quad [6.59]$$

This is the phenomenon that explains the abnormal heat capacity values of hydrogen (and deuterium) at low temperatures. At absolute zero, we only have the ortho form with symmetrical nuclear spins. By increasing the temperature, hydrogen gradually transforms into the para form with an anti-symmetrical function; the transformation would be complete at approximately 20 K. For all lower temperatures, in the presence of a catalyst, we obtain a mixture with variable proportions of the two forms of hydrogen.

In the case of non-linear polyatomic molecules, a number of symmetry σ will eventually be introduced by which we must divide the value given by relation [6.53]. This number will be equal to 1 if the molecule does not have a center of symmetry; it will be equal to 2 for water, 3 for NH_3 and 6 for benzene.

6.6.4. *Electronic partition function*

We usually only have to consider the lowest level, called the fundamental level, if ε_0 and ε_1 are the energies of the first two multiplicity levels g_0 and g_1, the electronic partition function can usually be written in the form of the sum of only two terms:

$$z_e = g_{0(e)} \exp\left(-\frac{\varepsilon_0}{k_B T}\right) + g_{1(e)} \exp\left(-\frac{\varepsilon_1}{k_B T}\right) \qquad [6.60]$$

Frequently the difference $\varepsilon_1 - \varepsilon_0$ is very large with regard to $k_B T$ such that the second term is negligible with regard to the first and can be written:

$$z_e = g_{0(e)} \exp\left(-\frac{\varepsilon_0}{k_B T}\right) \qquad [6.61]$$

It is worth choosing the energy ε_0 for the origin of electronic energies so that the electronic partition function is reduced to:

$$z_e = g_{0(e)} \qquad [6.62]$$

6.7. Partition function of an ideal gas

We will identify the ideal gas with an ensemble of punctual molecules, i.e. dimensionless and without mutual interaction. As the molecules are punctual, the free volume offered to the translation of molecules is the total volume of the container V.

As the molecules are not interacting, the potential energy only comprises the boundary term, the partition function of the ideal gas can be written using [6.5] and [6.23]:

$$z_{pf} = z_n z_e z_v z_r z_{tpf} = V\left(\frac{2\pi m k_B T}{h^2}\right)^{3/2} z_n z_e z_v z_r \qquad [6.63]$$

By regrouping the product of internal contributions due to rotational, vibrational, electronic and nuclear degrees of freedom under the term z_{int}, we obtain:

$$z_{pf} = V\left(\frac{2\pi m k_B T}{h^2}\right)^{3/2} z_{\text{int}} \qquad [6.64]$$

We often use the following form of this relation:

$$\frac{z_{pf}}{V} = \left(\frac{2\pi m k_B T}{h^2}\right)^{3/2} z_{\text{int}} \qquad [6.65]$$

This equation represents the contribution to the partition function of the ideal gas that does not take into account the potential energy of interaction with the boundaries of the container, that is to say that it only includes the kinetic energy of molecules in the translation. This external contribution will be identical for non-punctual molecules that interact with one another and with the boundaries, i.e. for real molecules.

6.8. Average energy and equipartition of energy

Now we will move onto indistinguishable particles but in the approximation of the classical limiting case. The canonical partition function is given by expression [5.25] using the molecular partition functions. By applying equation [5.32], we calculate for the internal energy:

$$U - U(0) = -\frac{1}{Z_C}\left(\frac{\partial \ln Z_C}{\partial \beta}\right) = -\frac{N}{z}\frac{\partial z}{\partial \beta} \qquad [6.66]$$

In the case where the different energetic contributions are independent and by applying expression [6.5] we can write:

$$U - U(0) = k_B T^2 \frac{N}{(z_t z_r z_v z_e)} \frac{\partial (z_t z_r z_v z_e)}{\partial T} \qquad [6.67]$$

Let $\langle \varepsilon_t \rangle$ be the average translational energy of molecules; as the internal energy is the sum of all the mean energies of the different modes, we can write:

$$\begin{aligned} U - U(0) &= N\{\langle \varepsilon_t \rangle + \langle \varepsilon_r \rangle + \langle \varepsilon_v \rangle + \langle \varepsilon_e \rangle\} \\ &= \{\langle E_t \rangle + \langle E_r \rangle + \langle E_v \rangle + \langle E_e \rangle\} \end{aligned} \qquad [6.68]$$

We must calculate the different contributions of this sum by obeying, for contribution m, the following relation, taking into account equation [6.67]:

$$\langle \varepsilon_m \rangle = -\frac{1}{z_m}\left(\frac{\partial z_m}{\partial \beta}\right) = \frac{k_B T^2}{z_m}\left(\frac{\partial z_m}{\partial T}\right) \qquad [6.69]$$

6.8.1. *Mean translational energy*

We must use the expression for the translational energy given for a three-dimensional system using equation [6.20]; by applying [6.69] (with N=N$_a$) it becomes:

$$\langle E_t \rangle = N_a \langle \varepsilon_t \rangle = RT^2 \frac{\Lambda^3}{x_1 x_2 x_3}\left(\frac{\partial\left(\frac{x_1 x_2 x_3}{\Lambda^3}\right)}{\partial T}\right) \qquad [6.70]$$

Referring again to relation [6.21], we obtain:

$$\langle \varepsilon_t \rangle = \frac{3}{2}\left(\frac{1}{\beta}\right) = \frac{3}{2} k_B T \qquad [6.71]$$

This expression is consistent with the equipartition theory of energy, which states that the contribution to the average energy of each quadratic term is $(1/2)k_B T$. In fact, the translational energy that in classical mechanics is in the form:

$$E_t = \frac{1}{2} m v_{x_1}^2 + \frac{1}{2} m v_{x_2}^2 + \frac{1}{2} m v_{x_3}^2 \qquad [6.72]$$

has three quadratic terms, each attached to one dimension.

The same calculation carried out in a box with only one dimension (x_1) would effectively give the value of a contribution: $(1/2)k_B T$.

Note that this mean energy does not depend on the dimensions of the container x_1, x_2 and x_3.

6.8.2. *Mean rotational energy*

Here, we are only dealing with a linear molecule with two degrees of freedom of rotation. At a high enough temperature ($\Theta_r \ll T$), the partition function is given by relation [6.51]. The simple application of equation [6.69] to the rotation immediately gives:

$$\langle \varepsilon_r \rangle = \mathrm{k_B}T \qquad [6.73]$$

The energy of a two-dimensional rotor with two quadratic terms in classical mechanics has the form:

$$E_r = \frac{1}{2} I\omega_{x_1}^2 + \frac{1}{2} I\omega_{x_2}^2 \qquad [6.74]$$

We thus find the equidistribution theory of energy with the contribution $(1/2)\mathrm{k_B}T$ of each quadratic term.

6.8.3. *Mean vibrational energy*

We will take the example of a harmonic oscillator with a single vibrational degree (diatomic molecule). According to relations [6.40] and [6.42], the corresponding partition function, taking the fundamental energy level (v=1/2) as the origin of energies (numerator of equation [6.42] is equal to1), is:

$$z_v = \frac{1}{1-\exp(-\mathrm{h}\nu\beta)} \qquad [6.75]$$

Applying relation [6.69] to the vibrational contribution gives:

$$\langle \varepsilon_v \rangle = \frac{\mathrm{h}\nu}{\exp\left(\dfrac{\mathrm{h}\nu}{\mathrm{k_B}T}\right)-1} \qquad [6.76]$$

If the temperature is quite high ($\Theta_v << T$) or ($h\nu\beta << 1$), we can develop the exponential term of the previous expression, and by keeping only the first non-zero term of the denominator, it becomes:

$$\langle \varepsilon_v \rangle \cong k_B T \qquad [6.77]$$

In classical mechanics, the energy of the harmonic oscillator is written:

$$E_r = \frac{1}{2} m v_{x_1}^2 + \frac{1}{2} k x_1^2 \qquad [6.78]$$

This expression has two quadratic terms and we can once again justify the equidistribution theory of energy with the contribution $(1/2) k_B T$ of each quadratic term.

6.9. Translational partition function and quantum mechanics

We calculated the partition functions of different movements using the energy levels given by quantum mechanics except for translation for which we used the energetic terms of classical mechanics by weighting them using a related degeneracy, that of Heisenberg's principle (equation [6.10]). We are going to show that the choice that helped us introduce interactions into section 6.4.2 is equivalent to using quantum mechanics, which we will show for translation with a single boundary effect.

Let us consider a gas molecule with mass m free to displace in a container with the volume Nv_f available for the displacement of the molecule. Let X_1, X_2 and X_3 be the length, width and height of the container:

$$Nv_f = X_1 X_2 X_3 \qquad [6.79]$$

Let ε_i be the sum of translational energies in the three directions, i.e:

$$\varepsilon_i = \varepsilon_{n_1}(X_1) + \varepsilon_{n_2}(X_2) + \varepsilon_{n_3}(X_3) \qquad [6.80]$$

n_1, n_2 and n_3 being, respectively, the quantum numbers for the movement in the directions Ox_1, Ox_2 and Ox_3. The translational partition function will therefore be:

$$z_t = \sum_{n_1, n_2, n_3} \exp\left(-\beta\varepsilon_{n_1} - \beta\varepsilon_{n_2} - \beta\varepsilon_{n_3}\right) \qquad [6.81]$$

or

$$z_t = \sum_{n_1} \exp\left(-\beta\varepsilon_{n_1}\right) \sum_{n_2} \exp\left(-\beta\varepsilon_{n_2}\right) \sum_{n_3} \exp\left(-\beta\varepsilon_{n_3}\right) \qquad [6.82]$$

which can be written in the form:

$$z_t = z_{t(x_1)} z_{t(x_2)} z_{t(x_3)} \qquad [6.83]$$

Quantum mechanics is used to calculate the energy levels of a molecule with mass m in a recipient with length X_1, i.e.:

$$\varepsilon_n = \frac{n^2 \mathrm{h}^2}{8mX_1^2} \qquad [6.84]$$

The lowest level has the energy:

$$\varepsilon_n = \left(n^2 - 1\right)\varepsilon \text{ with } \varepsilon = \frac{\mathrm{h}^2}{8mX_1^2} \qquad [6.85]$$

With the contribution of direction Ox_1 to the translational function:

$$z_{t(x_1)} = \sum_{n=1}^{\infty} \exp\left(-(n^2 - 1)\beta\varepsilon\right) \qquad [6.86]$$

With the translational energies being very narrow in a typical laboratory container, we can replace the equation with an integral and write the contribution according to Ox_1 in the form:

$$z_{t(x_1)} = \int_1^{\infty} \exp\left(-(n^2 - 1)\beta\varepsilon\right) \mathrm{d}n$$

By disregarding 1 with regard to n^2, through which we can change the variable $u^2 = n^2\beta\varepsilon$, we obtain:

$$z_{t(x_1)} = \left(\sqrt{\frac{1}{\beta\varepsilon}}\right) \int_1^{\infty} \exp\left(-(u^2\right) \mathrm{d}u = \sqrt{\frac{\pi}{4\beta\varepsilon}} \qquad [6.87]$$

which gives, by taking into account relation [6.85], for the contribution of direction Ox_1:

$$z_{t(x_1)} = \left(\frac{2\pi m k_B T}{h^2} \right)^{1/2} X_1 \qquad [6.88]$$

The complete translational partition function for the three directions will therefore be:

$$z_t = \left(\frac{2\pi m k_B T}{h^2} \right)^{3/2} X_1 X_2 X_3 = N v_f \left(\frac{2\pi m k_B T}{h^2} \right)^{3/2} \qquad [6.89]$$

This expression is identical to relation [6.19].

6.10. Interactions between species

After considering the data for each type of object, it is worth taking into account the interactions between them. The problem is modeling the system to deduce the interaction energies between objects. We will next consider these interactions between a certain number of objects, for example paired interactions, that is to say we will disregard any interaction between one object and another, except their immediate neighbors (which we used in section 6.4.2). The model can then become complex when considering interactions by triplets, quadruplets, etc.

When objects are *chemical* objects, we have two types of interaction: the interactions between charged particles (ions, electrons, etc.) and the interactions between neutral molecules based on dipole-dipole interactions.

6.10.1. *Interactions between charged particles*

Interactions between charged particles are due to Coulomb electrostatic forces that derive from a potential.

We can consider two simple interaction models between ions or more generally charged particles.

6.10.1.1. *Pairing interaction model*

We will only consider the interaction energy between two ions, immediate neighbors only, and we know that, for example, the interaction energy ε_{kj} between two spherical particles with radii a_j and a_k and respective charges z_k and z_j , separated from one another by $r_{j,k}$ (Figure 6.1), are in a constant dielectric environment ε such that:

$$\varepsilon_{kj} = \frac{|z_k||z_j|e^2}{4\pi\varepsilon r_{kj}} \text{ for } r_{kj} \geq a_j + a_k \quad [6.90]$$

and

$$\varepsilon_{kj} = \infty \text{ for } r_{kj} < a_j + a_k \quad [6.91]$$

This condition shows that the centers of the two ions cannot approach another by a distance less than a_j+a_k

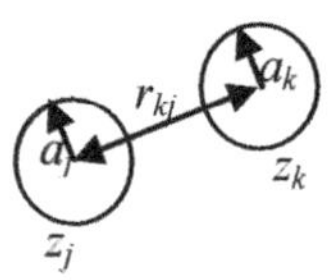

Figure 6.1. *Interaction between two ions*

6.10.1.2. *Ionic atmosphere*

The interaction model will take into account the ensemble of all ions of a medium and a determined ion is found in the field created by an ensemble of all others ions. This field will be calculated using Poisson's equation in a determined distribution or an average distribution of ions in the space. It is this mean that is used in the Debye-Hückel model of an ionic solution.

6.10.2. *Interaction energy between two neutral molecules*

Between two neutral models, three interaction models are used in the order of their complexity; the hard sphere model without force of attraction, the hard sphere model without repulsion and the London model for Van der Waals forces.

6.10.2.1. *The hard sphere model without force of interaction*

In this simple model, each molecule is supposedly protected by a sphere with diameter D and another molecule cannot enter this (Figure 6.2a); this gives function $\varepsilon(r)$, which is similar to that given by Figure 6.2b, which is defined by:

$$\begin{cases} \varepsilon(r) = +\infty & \textbf{if } r < D \\ \varepsilon(r) = 0 & \textbf{if } r \geq D \end{cases}$$

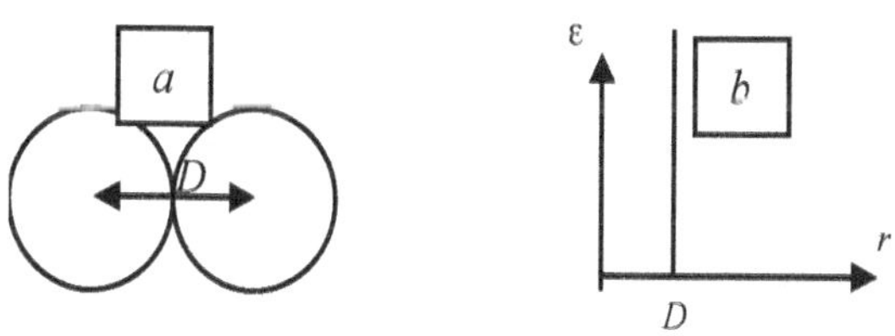

Figure 6.2. *The hard sphere model without attraction*

6.10.2.2. *The hard sphere model without Keesom repulsion force*

In this model, outside the hard protection sphere, there is an attraction energy of $1/r^6$ (as we will see in section 6.10.2.3 concerning the dipole-dipole origin); the function $\varepsilon(r)$ is represented in Figure 6.3 and is defined by, if ε_0 is the negative value of the mutual energy of two spheres protected by two molecules in contact with one another:

$$\begin{cases} \varepsilon(r) = +\infty & \textbf{if } r < D \\ \varepsilon(r) = \varepsilon_0 \left(\dfrac{D}{r}\right)^6 & \textbf{if } r \geq D \end{cases}$$

This definition gives rise to a molecular diameter D, which acts as the contact diameter.

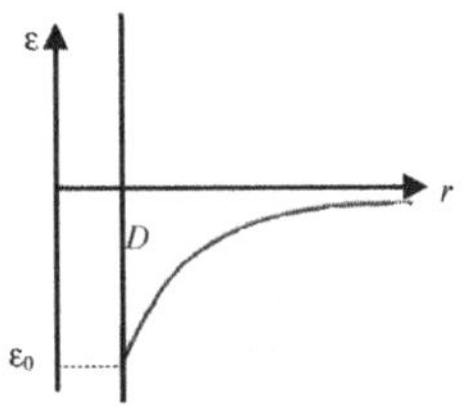

Figure 6.3. *Hard sphere model with force of attraction*

6.10.2.3. *The van der Waals force model*

In this model, the forces of interaction of two immobile molecules, called van der Waals forces, are represented as the sum of two terms: one of attraction and one of repulsion, each being a function of distance r, which separates the centers of the supposedly spherical molecules such that the first is predominant over a long distance and the second predominant over a short distance.

Now let us examine the term of attraction. The force of attraction is due to three different physical effects: the Keesom orientation effect, the Debye induction effect and the London dispersion effect.

The orientation effect is due to the fact that a certain number of molecules have a permanent dipolar electric moment due to the distribution of charges within the molecules. These electric dipoles orientate themselves according to one another and we can show that if molecules have moments $\vec{\mu}_1$ and $\vec{\mu}_2$, if the temperature is not too high (in practice if $\vec{\mu}_1 \vec{\mu}_2 << 1/\mathrm{k_B}T$), the corresponding energy of attraction is in the form:

$$\varepsilon_0 = -\frac{2\vec{\mu}_1^{2}\vec{\mu}_2^{2}}{3r^6}\frac{1}{\mathrm{k_B}T} \qquad [6.92]$$

The induction effect is due to the fact that, under the action of an electric field created by the permanent dipole of a molecule, another molecule is polarized, that is to say that the charge distributionin the second molecule is modified, which creates a dipole, called the induced dipole, whose moment is the product of the electric field by the polarizability of molecule α , which is a constant for this molecule, i.e.:

$$\vec{\mu} = \alpha \vec{E} \quad [6.93]$$

This induced moment is added to the permanent electric moment of the second molecule and Debye showed that the interaction energy between two molecules, due to this polarizability effect, is independent of temperature and takes the form:

$$\varepsilon_{\mathbf{in}} = -\frac{\alpha_{\mathbf{A}} \vec{\mu}_{\mathbf{B}}^{2} + \alpha_{\mathbf{B}} \vec{\mu}_{\mathbf{A}}^{2}}{r^6} \quad [6.94]$$

If molecules A and B are identical, the induction energy becomes:

$$\varepsilon_{\mathrm{in}} = -\frac{2\alpha\mu^2}{r^6} \quad \text{with } \alpha_{\mathrm{A}} = \alpha_{\mathrm{B}} = \alpha \text{ and } \bar{\mu}_{\mathrm{B}} = \bar{\mu}_{\mathrm{A}} = \bar{\mu} \quad [6.95]$$

The London dispersion effect is due to the following phenomenon. If at a given instant we could take a photo of a molecule, it should show, even in the absence of a permanent dipolar moment, an instantaneous dipolar moment. Even if the average of these instantaneous moments is zero, this instantaneous moment creates through the induction of a neighboring molecule, an induced instantaneous moment and London showed that initially the term of attraction due to this effect was in the form:

$$\varepsilon_{\mathrm{d}} = -\frac{3\mathrm{h}}{2r^6} \alpha_{\mathrm{A}} \alpha_{\mathrm{B}} \frac{\nu_{\mathrm{A}} \nu_{\mathrm{B}}}{\nu_{\mathrm{A}} + \nu_{\mathrm{B}}} \quad [6.96]$$

In this relation ν_{A} and ν_{B} are the frequencies of oscillation of the molecule, α_{A} and α_{B} are always the polarizabilities of molecules A and B.

If both molecules are identical, the dispersion energy takes the form:

$$\varepsilon_{\mathrm{d}} = -\frac{3\mathrm{h}\nu\alpha^2}{4r^6} \quad [6.97]$$

Note that the attraction energy is the sum of the three contributions [6.92], [6.94] and [6.97]. Given that the three terms vary in $1/r^6$, we can write:

$$\varepsilon_{attr} = \varepsilon_o + \varepsilon_I + \varepsilon_d = -\frac{m_0}{r^6} \qquad [6.98]$$

If we compare the energy levels due to each of the three effects, we find that contrary to what is commonly thought, it is the dispersion effect that is often greatest and that the contribution of the orientation effect present in dipolar molecules is often minor.

Now let us examine the repulsion term that should increase rapidly over a short distance due to the inter-penetrability of molecules. Although wave mechanics calculations give exponentially increasing potentials, we usually use a power law often with power 12, i.e. a repulsion potential in the form:

$$\varepsilon_{rep} = \frac{l_0}{r^{12}} \qquad [6.99]$$

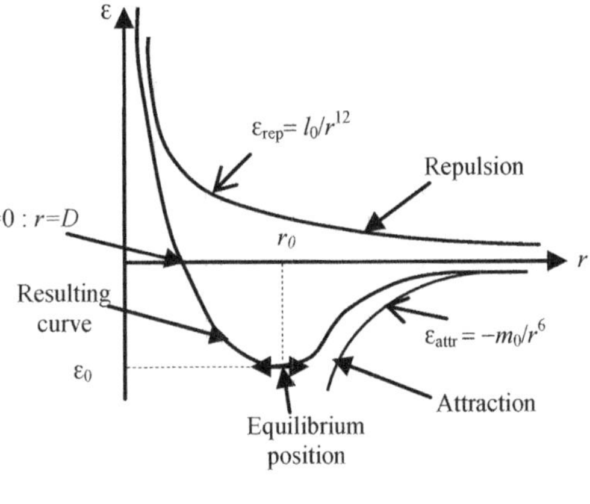

Figure 6.4. *London molecular interaction potential*

So, by combining the two expressions [6.98] and [6.99], we obtain the expression of the interaction potential, called the Lennard-Jones potential:

$$\varepsilon = -\frac{m_0}{r^6} + \frac{l_0}{r^{12}} \qquad [6.100]$$

Figure 6.2 shows the attraction potential, repulsion potential and the potential resulting from both effects as a function of the distance between the two molecules.

The resulting curve has a minimum ε_0 (<0) for a certain distance r_0 and the potential energy is canceled out for $r = D$. The minimum ε_0 is approximately -10^{-21} joules; r_0 is the angström order of magnitude, i.e. 10^{-10} m.

Considering this data, the coefficients used in relation [6.98] take the form:

$$m_0 = -2\varepsilon_0(r_0)^6 \qquad [6.101]$$

$$l_0 = -\varepsilon_0(r_0)^{12} \qquad [6.102]$$

The order of magnitude m_0 is 10^{-77} j.m^6; that of l_0 is approximately 10^{-138} J.m^{12}.

Relation [6.100] then becomes:

$$\frac{\varepsilon}{\varepsilon_0} = -2\left(\frac{r_0}{r}\right)^6 + \left(\frac{r_0}{r}\right)^{12} \qquad [6.103]$$

If D is the distance at which the energy is cancelled out (Figure 6.4), we can easily calculate:

$$\left(\frac{D}{r_0}\right)^6 = \frac{1}{2} \qquad [6.104]$$

The diameter D can still be considered as a true collision radius for infinitely slow molecules.

6.11. Equilibrium constants and molecular partition functions

The fact that thermodynamic constants and, in particular, Gibbs energies can be expressed using partition functions, this must be the same for equilibrium constants.

We are going to express the equilibrium constant of a reaction using the partition functions of the reactants and products.

A chemical reaction between reactants A_r giving products A_p can be written in the form:

$$0 = \sum_p a_p A_p + \sum_r a_r A_r = \sum_i a_i A_i$$

This equation uses stoichiometric algebraic numbers a_i, positive for a reaction product and negative for a reactant.

According to relation [5.46] and, by taking into account the expression of pressure given by relation [5.44], we can write the Gibbs energy of component i in the form:

$$G_i(T) - G_i(0) = -k_B T \ln Z_{iC} + P_i V \qquad [6.105]$$

$G_i(0)$ is the Gibbs energy of the pure compound i at a temperature of 0K.

We are going to consider three types of equilibrium: gaseous phase homogeneous, liquid phase homogenous and solid phase homogenous equilibria.

6.11.1. *Gaseous phase homogeneous equilibria*

Gas molecules are considered as indistinguishable molecules; the canonical partition function is therefore given by relation [5.25]. By applying this relation and by using Stirling's first approximation [4.1], relation [6.105] becomes:

$$G_i(T) - G_i(0) = -N_i k_B T \ln z_i + k_B T(N_i \ln N_i - N_i) + P_i V \qquad [6.106]$$

We are going to use molar variables. For this, we define a molar partition fraction $z_{i(m)}$ given by:

$$z_{i(m)} = \frac{z_i}{n_i} \qquad [6.107]$$

In the calculation of the translational partition function, this definition replaces the volume V by the molar volume $v_{i(m)}$.

Under standard conditions, at temperature T, the Gibbs energy in n moles of component i, which is considered to be an ideal gas ($P_iV = n_i\mathrm{R}T$), becomes:

$$g_i^0(T) = g_i^0(0) - n_i\mathrm{R}T \ln \frac{z_{i(m)}}{\mathrm{N_a}} \qquad [6.108]$$

For the reaction, we will write:

$$\Delta_r g_i^0(T) = \sum_i a_i g_i^0(T) = \sum_i a_i g_i^0(0) - \mathrm{R}T\sum_i a_i \ln \frac{z_{i(m)}}{\mathrm{N_a}} \qquad [6.109]$$

However, at a temperature of 0K, for an ideal gas, we obtain:

$$g_i^0(0) = u_i^0(0) + PV - TS = u_i^0(0) + \mathrm{R}T - TS = u_i^0(0) = h_i^0(0) \qquad [6.110]$$

Consequently, relation [6.109] becomes:

$$\Delta_r g_i^0(T) = \Delta_r u^0(T) - \mathrm{R}T\sum_i a_i \ln \frac{z_{i(m)}}{\mathrm{N_a}} \qquad [6.111]$$

The term $\Delta_r u^0(T)$ is the linear combination, weighted by stoichiometric numbers, of the residual vibrational energies of each substance. Therefore if each component has k_i degrees of vibrational freedom with the fundamental frequency $\nu_{k_i}^0$, we will obtain:

$$\Delta_r u^0(0) - \frac{\mathrm{N_a h}}{2}\sum_i (a_i \sum_k \nu_{k_i}^0) = \Delta_r h^0(0) \qquad [6.112]$$

The equilibrium constant is defined by:

$$-\mathrm{R}T \ln K_P = \Delta_r g^0(T) \qquad [6.113]$$

Taking into account relation [6.111], we obtain:

$$K_P = \prod_{i} \left(\frac{z_{i(m)}}{N_a} \right)^{a_i} \exp - \frac{\Delta_r h^0(0)}{RT} \qquad [6.114]$$

This relation is valid; let us note this for gas-phase equilibria, when ideal gases are involved.

6.11.2. *Liquid phase homogeneous equilibria*

If the reaction occurs in a liquid phase then all reaction components are therefore components of the same liquid solution.

For such a solution, the change in volume due to the reaction is negligible and we obtain:

$$\Delta_r G(T) = \Delta_r F(T) + P\Delta_r V \cong \Delta_r F(T) \qquad [6.115]$$

We can express the Gibbs energy using relation [5.42]. As gas molecules are considered as indistinguishable molecules, the canonical partition is therefore given by relation [5.25]. Using Stirling's second approximation [4.1], relation [6.115] becomes:

$$F_i(T) - F_i(0) = -N_i k_B T \left(\ln z_i + \ln N_i \right) \qquad [6.116]$$

In this expression, we assume that the solution is ideal as it does not take into account the enthalpy of mixing.

Under standard conditions, at temperature T, the Gibbs energy of n moles of component i, becomes:

$$f_i^0(T) = f_i^0(0) - n_i RT \ln \frac{z_{i(m)}}{N_a} \qquad [6.117]$$

For the reaction, we obtain:

$$\Delta_r f_i^0(T) = \sum_i a_i f_i^0(0) = -RT \sum_i a_i \ln \frac{z_{i(m)}}{N_a} \qquad [6.118]$$

$f_i^0(0)$ is the standard Gibbs molar energy of pure component i at a temperature of 0 K; $z_{i(m)}$ is the molar partition function.

Using relations [6.113] and [6.116], the equilibrium constant can be written:

$$-RT\ln K_x = \Delta_r f^0(T) \qquad [6.119]$$

Using relation [6.118], we obtain:

$$-RT\ln K_x = \Delta_r\left[h_i^0(0)\right] - RT\ln\prod_i\left(\frac{z_{i(m)}}{\mathrm{N_a}}\right)^{a_i} \qquad [6.120]$$

The equilibrium constant becomes:

$$K_x = \prod_i\left(\frac{z_{i(m)}}{\mathrm{N_a}}\right)^{a_i}\exp\left(-\frac{\Delta_r h_0}{RT}\right) \qquad [6.121a]$$

This relation, which is valid in the case of ideal or very dilute solutions, is in the same form as relation [6.114] obtained for a homogeneous reaction between ideal gases.

Some authors introduce the molecular partition function, defined by:

$$z_{i(m)}^* = \frac{z_{i(m)}}{\mathrm{N_a}} \qquad [6.121b]$$

Equality [6.121a] then takes the form:

$$K_x = \prod_i\left(z_{i(m)}^*\right)^{a_i}\exp\left(-\frac{\Delta_r h_0}{RT}\right) \qquad [6.122]$$

For solutions with deviations from ideality, we must add the enthalpy of mixing to the zero enthalpy, calculated by relation:

$$\Delta_r H^{mix} = \sum_i a_i \overline{H}_i^{mix} \qquad [6.123]$$

Relation [6.121] then takes the form:

$$K_x = \prod_i \left(z_{i(m)}^*\right)^{a_i} \exp\left(-\frac{\Delta_r h_0 + \Delta_r H^{mix}}{RT}\right) \qquad [6.124]$$

6.11.3. *Solid phase homogenous equilibria*

If the reaction proceeds in a solid phase, then all the reaction components are components of the same solid solution.

For such a solution, as with liquids, the change in volume due to the reaction is negligible and we obtain:

$$\Delta_r G(T) = \Delta_r F(T) + P\Delta_r V \cong \Delta_r F(T) \qquad [6.125]$$

We can express the Gibbs energy by relation [5.42]. The molecules of the solid are therefore considered as distinguishable molecules; the canonical partition function is therefore given by relation [5.15]. So relation [6.125] becomes:

$$F_i(T) - F_i(0) = -N_i \mathrm{k_B} T \ln z_i \qquad [6.126]$$

In this expression, we assume the solution to be ideal as it does not take into account the enthalpy of mixing.

Under standard conditions, at temperature T, the Gibbs energy in n moles of component i, becomes:

$$f_i^0(T) = f_i^0(0) - n_i \mathrm{R} T \ln z_i \qquad [6.127]$$

NOTE.– In the case of solids, there is no need to include molar partition functions, as the partition functions of solids only contain vibrational terms, which do not include the volume of the container.

For the reaction, we can write:

$$\Delta_r f_i^0(T) = \sum_i a_i f_i^0(0) = -RT\sum_i a_i \ln z_i \tag{6.128}$$

$f_i^0(0)$ is the standard molar Gibbs energy of the pure component i at a temperature of 0 K; z is its function partition.

Using relations [6.113] and [6.126], the equilibrium constant is written:

$$-RT \ln K_x = \Delta_r f^0(T) \tag{6.129}$$

Using relation [6.128], we obtain:

$$-RT \ln K_x = \Delta_r \left[h_i^0(0)\right] - RT \ln \prod_i (z_i)^{a_i} \tag{6.130}$$

The equilibrium constant is therefore:

$$K_x = \prod_i (z_i)^{a_i} \exp\left(-\frac{\Delta_r h_0}{RT}\right) \tag{6.131}$$

6.12. Conclusion on the macroscopic modeling of phases

Therefore, by defining the phase as the collection of a large number of objects, statistical thermodynamics helps to calculate a variable called the canonical partition function which links:

– on the one hand, molecular or microcanonical partition functions calculated from the characteristic of objects (masses, vibrational frequencies, moments inertia);

– on the other hand, classical potential thermodynamic functions defined in Chapter 1.

Modeling therefore becomes easier in the case of systems without interactions between molecules. In practice, the main difficulty found with this modelling is the need to introduce interaction forces between the objects in the collection.

7

Pure Real Gases

In this chapter, we will focus on a system with two variables, either V and T or P and T. We know that to describe a pure compound we must know either a characteristic function (F or G), or three thermodynamic coefficients: two diagonal terms of a characteristic matrix and a third term; the fourth is deduced from the symmetry of the characteristic matrix.

7.1. The three states of the pure compound: critical point

We know that a pure substance can exist, within normal temperature ranges, in three states: gas, liquid and solid. A stable system containing a pure compound can be formed from one, two or three phases in equilibria. These three phases are commonly represented in a temperature–pressure diagram such as that in Figure 7.1. This diagram shows three lines representing the change of state separating the monophasic region. The sublimation line separates the stability zone of a solid and that of a gas; the melting line separates the liquid and solid zone and the vaporization line separates the liquid zone from the gas zone. There are also two interesting points: the triple point where the three phases are in equilibria and the critical point K. Note that this point is the point of maximal temperature and pressure, where a pure compound can exist in a liquid–gas equilibrium. Above K, the distinction between a liquid and a gas disappears, the meniscus that separates the two phases also disappears and if we move about the critical point from A to B, it is possible to pass from a liquid state in A, where the two phases are separated by a meniscus, into a gaseous state in B, without crossing the limit of the phase. The zone situated at a higher temperature and pressure than the critical temperature is sometimes called the *supercritical fluid* zone.

NOTE.– We distinguish a gas, which is the phase present when the system contains a pure compound in the gaseous state only, from a vapor, which also qualifies as a gaseous phase, when the system contains both gaseous and liquid states.

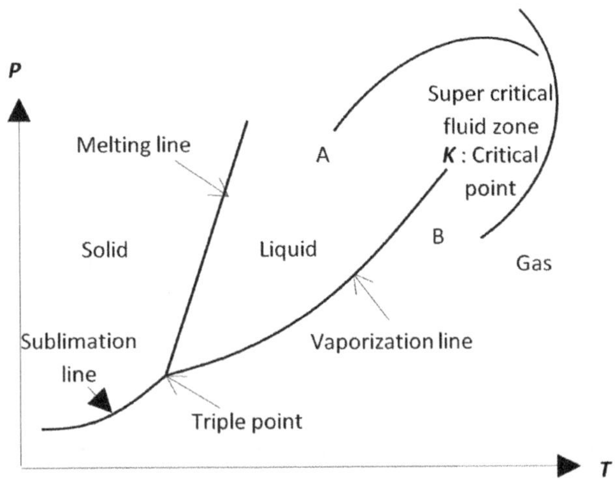

Figure 7.1. *Main lines in the diagram of pure compound states*

In this chapter, we are going to study the properties of pure gaseous phases.

7.2. Standard state of a molecular substance

We define the standard state of a molecular substance at temperature T as the state of a pure substance at this temperature, under 1 bar of pressure, in the aggregation state that it has naturally under these conditions.

If it is a solid, it will be of a stable crystallographic variety. For example, the standard state of carbon at 25°C is graphite.

Liquids, as well as gases, are considered as being in the same aggregation state.

In the standard state, the behavior of a gas will be considered as that of a perfect gas, even if at a pressure of 1 bar and at the chosen temperature, the actual behavior of the gas is not the same as the behavior of a perfect gas.

7.3. Real gas – macroscopic description

A real gas has a behavior that differs from that of a perfect gas. This is clearly visible if we plot the product PV according to pressure at different temperatures (Figure 7.2).

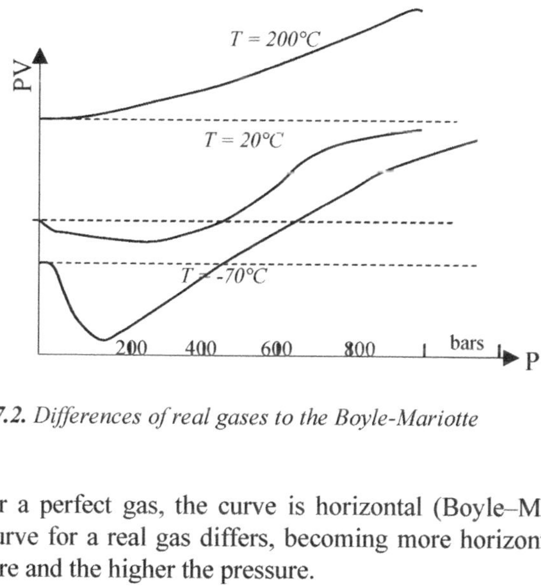

Figure 7.2. Differences of real gases to the Boyle-Mariotte

We find that for a perfect gas, the curve is horizontal (Boyle–Mariotte law) whereas the curve for a real gas differs, becoming more horizontal the lower the temperature and the higher the pressure.

Usually, a gas is macroscopically defined either by a diagram $(P - V)$ or $(PV\text{-}P)$ and a heat capacity or by its state equation $f(P, V, T, n) = 0$ and a heat capacity. The thermodynamic coefficients can therefore easily be calculated either graphically or algebraically.

7.3.1. *Pure gas diagram (P-V)*

The first sources of data for the pure gas are the (P, V) diagrams of each gas (Figure 7.3(b)), which give pressure as a function of volume for one mole of gas at different temperatures. At high temperatures, this curve is a hyperbole given by the perfect gas law. Gradually, as the temperature decreases, the curves deform until they show an inflexion point to the horizontal tangent for the critical temperature (point K). The curve then takes the form of a saturation curve for the appearance of a liquid. Below the saturation curve, there is no longer an isotherm as the gas alone is not stable; hence, here we have a liquid.

These diagrams exist for many gases and can be digitized, replacing the curves with files (P, V, T) which can be used in computer calculations.

7.3.2. *"Cubic" state equations*

There is no universal state equation like the state equation for a perfect gas – each equation has its advantages but also its disadvantages. The more adjustable constants the equation has the better it represents a given gas; the best state equation does not require too many adjustment measurements, that is to say, one that contains a limited number of constants. All equations proposed tend towards the perfect gas equation since, as shown in Figure 7.2, a real gas tends towards a perfect gas when the pressure tends towards zero.

"Cubic" state equations are third degree equations with regards to volume. The most well-known of these equations is the Van der Waals equation.

These equations account for a critical point whose coordinates are obtained by canceling out the two primary partial derivatives of pressure with regard to volume.

Under critical conditions ($P<P_c$ and $T<T_c$), canceling out of the first derivative gives an equation that has either one or three roots.

If such a state equation gives a single root, this proves that the system is monophasic and the root gives the molar volume of this phase.

If such a state equation gives three roots, the lowest value of the molar volume obtained corresponds to the molar volume of the liquid and the highest value corresponds to the molar volume of the gas. The intermediate value has no physical sense and the $P(T)$ curve between the two extreme roots is replaced by a horizontal line, since it represents a diphasic system in a liquid–vapor equilibrium and in this case the pressure is no longer a function of volume, it is constant while the two phases are present.

7.3.2.1. *Van der Waals state equation*

The van der Waals equation is the most well-known cubic state equation for non-perfect gases, but also certainly one of the most mediocre. However, despite its downfalls, which limit its practical use, this equation is of high educational interest due to four of its qualities:

– it only contains two adjustable constants;

– it can be justified by a microscopic approach that gives some physical sense to its two constants;

– it includes the change in state from liquid-gas, which shows that this change in state is part of the description of the phase if artificially inducing the liquid-gas transition is usually only due to our ignorance. However, the van der Waals equation is undoubtedly still one of the best to represent liquids;

– finally, it takes into account the existence of a critical state.

The Van der Waals equation for one mole, if v^0 is the molar volume, is written:

$$\left(P+\frac{a}{\left(v^0\right)^2}\right)\left(v^0-b\right)=\mathrm{R}T \tag{7.1}$$

Or, in its extensive form for n moles in volume V:

$$\left(P+\frac{n^2 a}{V^2}\right)(V-nb)=n\mathrm{R}T \tag{7.2}$$

Parameter a is called the cohesion pressure, its order of magnitude is between 1 and 10^{-3} Pa.m^6.mole^{-2}. Parameter b is the co-volume whose order of magnitude is 10^{-5} m^3.mole^{-1}.

Figure 7.3(a) represents the surface of Van der Waals f(P, V, T) = 0. Figure 7.3(b) shows some isotherms.

The internal energy of a real gas, unlike that of a perfect gas, depends on volume; for example, in the case of the Van der Waals state equation, it depends on volume, and can be easily calculated:

$$\frac{\partial U_m}{\partial V} = \frac{a}{\left(v^0\right)^2} \qquad [7.3]$$

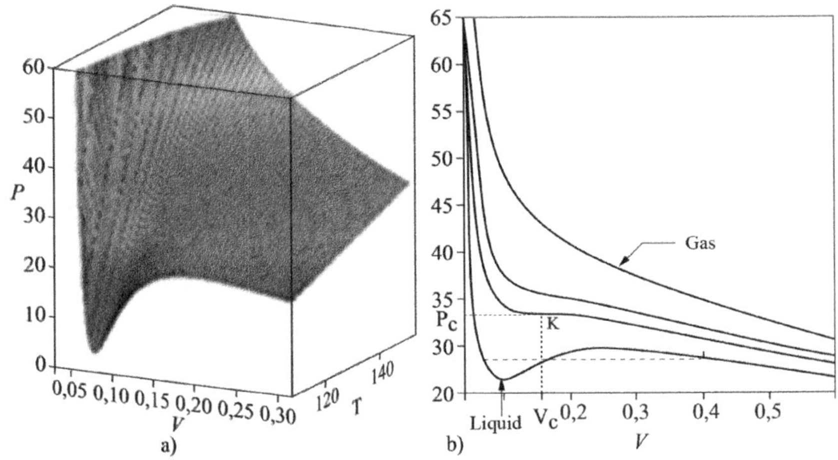

Figure 7.3. *(a) Representation of the van der Waals equation in the system of axes OVTP-; (b) representation of the isotherm P(V)*

On the isotherm with a single inflexion point to the tangent horizontal, we can write that the critical point corresponds to coordinates of this inflexion point (point K in Figure 7.3(b)): this is what we obtain mathematically, in addition to complying with equation [7.2]:

$$\frac{\partial P}{\partial V} = -\frac{\mathrm{R}T}{\left(V-b\right)^2} + \frac{2a}{V^3} = 0 \quad \text{and} \quad \frac{\partial^2 P}{\partial V^2} = \frac{2\mathrm{R}t}{\left(V-b\right)^3} - \frac{6a}{V^4} = 0 \qquad [7.4]$$

The expression of coordinates of the critical point can be deduced using two parameters, a and b:

$$T_c = \frac{8a}{27\mathrm{R}b}; \; P_c = \frac{a}{27b^2}; \; V_c = 3nb \qquad [7.5]$$

Building on the Van der Waals equation, we can obtain the Berthelot equation, which takes into account the influence of temperature on the Van der Waals coefficient a; this is still a cubic state equation with two parameters which is a bit closer to reality. For one mole of gas it is:

$$\left(P+\frac{a}{T\left(v^{0}\right)^{2}}\right)\left(v^{0}-b\right)=\mathrm{R}T \qquad [7.6]$$

Coefficient a of both relations [7.1] and [7.6] are clearly not identical.

7.3.2.2. *The RK state equation of Redlich and Kwong*

This equation – Redlich and Kwong (RK) – is much improved with regards to the Van der Waals equation but it is still unable to accurately describe the liquid state and therefore gas-liquid equilibria. For one mole, it takes the following form:

$$P=\frac{\mathrm{R}T}{v^{0}-b}-\frac{a}{v^{0}\left(v^{0}+b\right)\sqrt{T}} \qquad [7.7]$$

The different coefficients are given by:

$$a=\frac{0.42748\mathrm{R}^{2}T_{c}^{2,5}}{P_{c}}\ ;\ b=\frac{0.08662\mathrm{R}T_{c}}{P_{c}} \qquad [7.8]$$

This equation requires the condition:

$$\frac{P}{P_{c}}<\frac{T}{2T_{c}} \qquad [7.9]$$

7.3.2.3. *SRK state equation of Soave, Redlich and Kwong*

The RK equation was corrected by Soave, who proposed:

$$P=\frac{\mathrm{R}T}{v^{0}-b}-\frac{\alpha a}{v^{0}\left(v^{0}+b\right)} \qquad [7.10]$$

with the following coefficient values:

$$a = \frac{0.42748\ R^2 T_c^2}{P_c}\ ;\ b = \frac{0.0866 R T_c}{P_c} \qquad [7.11]$$

The coefficient α is given as a function of the relative temperature T_r, that is to say, the ratio between temperature at the critical temperature, by relation:

$$\alpha = \left[1 + \left(0.48 + 1.574\omega - 0.176\omega^2\right)\left(1 - T_r^{0.5}\right)\right]^2 \qquad [7.12]$$

ω is called the *coefficient of acentricity*. It was introduced by Pritzer in 1955 and take into account the non-sphericity of molecules; it is set to zero for monoatomic gases such as rare gases whose value is:

$$\omega = \log P_r^{sat} - 1 \text{ at } T_r = 0.7 \qquad [7.13]$$

P_r^{sat} is the pressure of relative saturation vapor, with regard to the pressure voltage at the critical pressure.

Graboski and Daubert proposed a modification of coefficient α given by relation [7.12] and propose:

$$\alpha = \left[1 + \left(0.48508 + 1.5547\omega - 0.15613\omega^2\right)\left(1 - T_r^{0.5}\right)\right]^2 \qquad [7.14]$$

Table 7.1 gives the value of the coefficient of acentricity of several gases.

Gas	ω	Gas	ω	Gas	ω
Neon	0	Chlorine	0.073	Methane	0.011
Argon	-0.004	Bromine	0.132	Ethylene	0.087
Krypton	-0.002	CO_2	0.223	Ethane	0.100
Xenon	0.002	CO	0.049	Benzene	0.212
Hydrogen	-0.22	NH_3	0.250	Toluene	0.257
Oxygen	0.021	HCl	0.12	*n*-Heptane	0.350
Nitrogen	0.037	H_2S	0.100	Propane	0.153
Fluorine	0.048	SO_2	0.251	*m*-Xylene	0.331

Table 7.1. *Coefficient of acentricity of several gases*

7.3.2.4. *Peng and Robinson state equation*

It is a variant of the previous equation which gives better results for the liquid phase and therefore for liquid-vapor equilibria:

$$P = \frac{\mathrm{R}T}{v^0 - b} - \frac{\alpha a}{\left(v^0\right)^2 - 2bv^0 - b^2} \quad [7.15]$$

with the coefficients defined by:

$$a = \frac{0.457235\ \mathrm{R}_2\mathrm{T}_c^2}{P_c};\ b = \frac{0.0777\mathrm{RT}_c}{P_c} \quad [7.16]$$

$$\alpha = \left[1 + \left(0.37464 + 1.54226\omega - 0.26992\omega^2\right)\left(1 - T_r^{0.5}\right)\right]^2 \quad [7.17]$$

7.3.3. ***Other state equations***

Besides *cubic* equations, the number of state equations is constantly increasing; of the different general laws proposed, we will only mention the most widespread.

7.3.3.1. *Dieterici equation*

Another form of the state equation with two parameters, but which is no longer a cubic equation, proposed by Dieterici, is written:

$$P\left(v^0 - b\right) = \mathrm{R}T \exp\left(-\frac{a}{\mathrm{R}Tv^0}\right) \quad [7.18]$$

which can be represented, by developing the exponential in series, as follows:

$$Pv^0 = \mathrm{R}T\left[1 + \left(b - \frac{a}{\mathrm{R}T}\right).\frac{1}{v^0} + \frac{a^2}{(\mathrm{R}T)^2}.\frac{1}{\left(v^0\right)^2} + \ldots\right] \quad [7.19]$$

7.3.3.2. Development of the virial

We find that the previous form of the Dieterici equation is a development in series of $1/v^0$. A more general form of such a development, called the virial equation, was proposed by Kamerlingh-Omnes and Keesom:

$$PV = n\mathrm{R}T\left[1+\frac{B_2}{V}+\frac{B_3}{V^2}+...\right] \quad [7.20]$$

Coefficients B_2, B_3, etc. are called the second, third, etc. coefficient of the virial. We can calculate the first coefficients of the virial using the cubic state equations as we did for the Dietrerici equation, but it does not provide any additional information. For example, for the Van der Waals equation we obtain, respectively:

$$Pv^0 = \mathrm{R}T\left[1+\left(b-\frac{a}{\mathrm{R}T}\right)\frac{1}{v^0}+\frac{ab}{\mathrm{R}T}\cdot\frac{1}{\left(v^0\right)^2}\right] \quad [7.21]$$

and for the Berthelot equation:

$$Pv^0 = \mathrm{R}T\left[1+\left(b-\frac{a}{\mathrm{R}T^2}\right)\frac{1}{v^0}+\frac{ab}{\mathrm{R}T^2}\cdot\frac{1}{\left(v^0\right)^2}\right] \quad [7.22]$$

Let ρ be the density of de molecules N/V, so that the virial equation [7.20] is written:

$$\frac{P}{\rho \mathrm{k}_B T} = 1+\frac{B_2}{V}+\frac{B_3}{V^2}+... \quad [7.23]$$

This series can be written as a MacLaurin series using $\rho = 0$, i.e.:

$$\frac{P}{\rho \mathrm{k}_B T} = 1+\frac{\partial}{\partial \rho}\left(\frac{P}{\rho \mathrm{k}_B T}\right)_{T,\rho=0}+\frac{1}{2}\frac{\partial^2}{\partial \rho^2}\left(\frac{P}{\rho \mathrm{k}_B T}\right)_{T,\rho=0}+... \quad [7.24]$$

So the coefficients of the virial are defined by:

$$B_i = \frac{N}{(i-1)!}\frac{\partial^{i-1}}{\partial\rho^{i-1}}\left(\frac{P}{\rho k_B T}\right)_{T,\rho=0} \quad [7.25]$$

It is therefore possible to measure the coefficients of the virial using pressure measurements at a constant volume for different values of the density ρ of the molecules.

NOTE.– The previous series of product Pv^0 can be replaced by a development of the same product according to the powers of pressure in the form:

$$Pv^0 = \mathrm{R}T\left[1 + B'_2 P + B'_3 P^2 + ...\right] \quad [7.26]$$

The main challenge is to give a physical sense to the different coefficients and especially to link them to the molecular properties of the system.

NOTE.– By making do with a corrective term of the perfect gas, the relation between the coefficients B_2 and B'_2, which respectively appear in relations [7.22] and [7.26], is:

$$B'_2 = B_2\mathrm{R}T \quad [7.27]$$

7.3.3.3. *Beattie–Bridgeman state equation*

One equation that is widely used, especially for gas mixtures, is called the *Beattie–Bridgeman equation.* With five constants (A_0, B_0, a, b and c), which greatly contributes to its reliability, it is written as:

$$P = \frac{\mathrm{R}T(1-\varepsilon)}{\left(v^0\right)^2}(v^0 - B) - \frac{A}{\left(v^0\right)^2} \quad [7.28]$$

with the following meanings:

$$A = A_0\left(1 - \frac{a}{v^0}\right), \; B = B_0\left(1 - \frac{b}{v^0}\right) \text{ and } \varepsilon = \frac{c}{v^0 T^2} \quad [7.29]$$

As above, the Beattie–Bridgeman equation can be written in the form of a virial, limited to four terms:

$$Pv^0 = \mathrm{R}T\left[1+\beta\frac{1}{v^0}+\gamma\frac{1}{\left(v^0\right)^2}+\delta\frac{1}{\left(v^0\right)^3}\right] \quad [7.30]$$

with the following meanings:

$$\beta = B_0 - \frac{A_0}{\mathrm{R}T} - \frac{c}{T^2}, \quad \gamma = -B_0 b + \frac{aA_0}{\mathrm{R}T} - \frac{B_0 c}{T^3} \text{ and } \delta = \frac{\mathrm{R}B - bc}{\mathrm{R}T^3}$$

7.3.4. *The theorem of corresponding states and the generalized compressibility chart*

The *reduced variables* (reduced pressure, reduce temperature, reduced molar volume) of a gas is defined by the ratios of the variable corresponding to the value at the critical point:

$$P_r = P/P_c, \; T_r = T/T_c \text{ and } V_r = v^0 / v_c \quad [7.31]$$

The *compressibility coefficient* is defined by:

$$Z = \frac{Pv^0}{\mathrm{R}T} \quad [7.32]$$

Reviewing the experimental data from the state charts allowed people to propose what we call the *theorem of corresponding states*. According to this principle, real gases with the same reduced volume exert the same reduced pressure at the same temperature. In addition, in the same reduced state, all gases will have the same compressibility factor

The term “principle” is not the most appropriate for this rule as this approximation is not rigorous; in fact, there are slight differences between different gases, which led Pitzer to introduce the corrective term, the acentric coefficient. Nevertheless, the rule is still a good approximation.

Compressibility charts (Figure 7.4) give the compressibility coefficient Z as a function of the reduced pressure for differences reduced temperatures. This result has a good acceptability for a certain number of gases.

Knowing the coordinates of the critical point of a gas, it is easy to calculate for each pressure and temperature its reduced coordinates. On the diagram in Figure 7.4, we choose the curve corresponding to the calculated reduced temperature and we determine the Z value for the chosen reduced pressure. Then using relation [7.32] we calculate its molar volume by relation:

$$v^0 = \frac{RT}{P} Z \qquad [7.33]$$

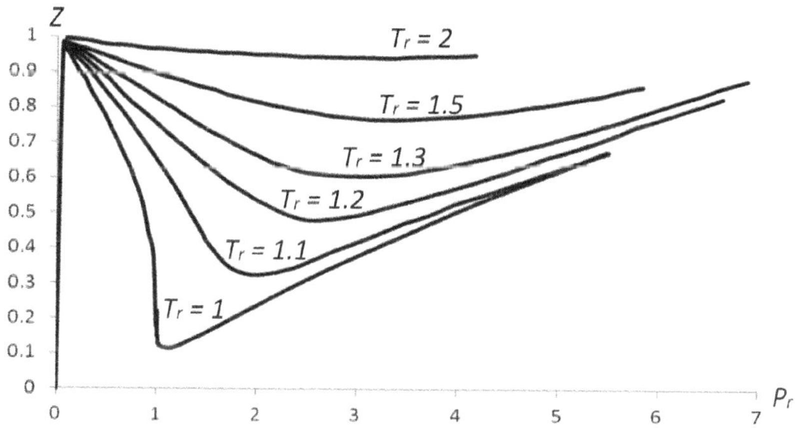

Figure 7.4. *Compressibility chart*

If the theorem of corresponding states is valid then the state equations for different gases should all be identical. This is, effectively, what we observe for the Van der Waals equation for which we substitute the reduced in equations [7.1] and [7.5]; we obtain a single equation:

$$\left(P_r + \frac{3}{V_r^2}\right)(3V_r - 1) = 8T_r \qquad [7.34]$$

We obtain the same type of result every time the state equation is a cubic equation, whose two parameters can be calculated from the two definition relations of the critical point, namely the canceling out of both derivatives

$\frac{\partial P}{\partial V}$ and $\frac{\partial^2 P}{\partial V^2}$. For equations with more than two parameters, the two conditions of criticality are not sufficient to determine these parameters.

7.3.5. *Molar Gibbs energy or chemical potential of a real gas*

Another way of defining a gas is to express a characteristic function. The most common example chosen is molar Gibbs energy, in the case of a pure gas, often interchangeable with chemical potential. Here we are focusing on the choice of variables P and T, and we aim to determine the chemical potential (or molar Gibbs energy) of the pure gas from the state equation written in the form:

$$v^0 = V(T,P) \qquad [7.35]$$

So the volume is the first derivative of Gibbs energy with regard to pressure (at a constant temperature):

$$v^0 = \left(\frac{\partial g^0}{\partial P}\right)_T \qquad [7.36]$$

By integration between pressures P^0 and P, it becomes:

$$g^0(T,P) = g^0(T,P^0) + \int_{P^0}^{P} V(T,P)dP \qquad [7.37]$$

If P^0 tends towards zero, we know that the gas tends towards the behaviour of a perfect gas, thus affecting the exponent * the variables relative to the perfect gas, it becomes:

$$g^0(T,P) = \lim_{P^0 \to 0} g^{0*}(T,P^0) + \lim_{P^0 \to 0} \int_{P^0}^{P} V(T,P)dP \qquad [7.38]$$

Often it is sufficient to choose, for the reference pressure, the pressure of 1 bar, which is low enough.

For a perfect gas in the same pressure domain, we obtain:

$$g^{0*}(T,P) = \lim_{P^0 \to 0} g^{0*}(T,P^0) + \lim_{P^0 \to 0} \int_{P^0}^{P} V^*(T,P)dP \qquad [7.39]$$

Now let g^* be the molar Gibbs energy of a perfect gas at the reference pressure and according to temperature:

$$g^{0*}(T,P) = g^*(T) + RT\ln\frac{P}{P^0} \qquad [7.40]$$

By subtracting the two relations [7.38] and [7.39], it becomes:

$$g^0(T,P) = g^{0*}(T,P) + \lim_{P^0 \to 0} \int_{P^0}^{P} \left(V(T,P) - V^*(T,P)\right)\mathrm{d}P \qquad [7.41]$$

Using relation [7.40] in expression [7.41], we obtain:

$$g^0(T,P) = g^*(T) + RT\ln\frac{P}{P^0} + \lim_{P^0 \to 0} \int_{P^0}^{P} \left(V(T,P) - \frac{RT}{P}\right)\mathrm{d}P \qquad [7.42]$$

Knowing the state equation or the variations in the compressibility coefficient Z with pressure, we can then calculate the integral of the second member and therefore molar Gibbs energy of the real gas. However, this method does not allow us to calculate the term $g^*(T)$ which proves the inability of a state equation to completely define a gas.

7.3.6. *Fugacity of a real gas*

7.3.6.1. *Definition*

The chemical potential of a perfect gas is given by relation [7.40]. Due to the simplicity of this formula, in particular when used to study physical or chemical equilibria, Lewis proposed an expression for a real gas by replacing the pressure by function $f^0(T,P)$, called the *fugacity*, and under these conditions, the chemical potential takes the following form, with a reference pressure P^0:

$$g^0(T,P) = g^*(T) + RT\ln\frac{f^0(T,P)}{P^0} \qquad [7.43]$$

Let us define the *fugacity coefficient* by:

$$\varphi^0 = \frac{f^0(P,T)}{P} \qquad [7.44]$$

For a perfect gas $\phi^0 = 1$, at low pressure, the gas tends to become perfect:

$$\lim_{P^0 \to 0} \frac{f^0(P,T)}{P} = \lim_{P^0 \to 0} \varphi^0 = 1 \qquad [7.45]$$

7.3.6.2. *Calculating the fugacity coefficient*

By comparing the equations [7.42], [7.43],[7.44] and [7.45], we obtain:

$$\mathrm{R}T \ln \varphi^0 = \lim_{P^0 \to 0} \int_{P^0}^{P} \left(V(P,T) - \frac{\mathrm{R}T}{P} \right) \mathrm{d}P \qquad [7.46]$$

This relation allows us to calculate the fugacity coefficient and therefore the fugacity from the state equation, for example.

NOTE.– By comparing expressions [7.41] and [7.46], we can write:

$$\mathrm{R}T \ln \varphi^0 = g^0(T,P) - g^0{}^*(T,P) \qquad [7.47]$$

7.3.6.3. *Approximate value of the fugacity coefficient*

Let us take definition [7.32] of the compressibility factor and express it using the virial relative to the pressures defined in relation [7.26], by being limited to the first terms, it becomes:

$$Z = 1 + \frac{B'_2 P}{\mathrm{R}T} \qquad [7.48]$$

Relation [7.46] can be written in the form:

$$\ln \varphi_0 = \frac{B'_2 P}{\mathrm{R}T} \approx Z - 1 \qquad [7.49]$$

If the fugacity coefficient is not too far from 1, we can write:

$$\ln \varphi^0 \approx \varphi^0 - 1 \qquad [7.50]$$

So, by comparing [7.49] and [7.50], we obtain:

$$\varphi^0 \approx Z \qquad [7.51]$$

This approximate value avoids the need to integrate relation [7.46] while providing good results. For example, up to pressures of 100 atm, at not too high temperatures, close to ambient pressure, we obtain a result correct for oxygen and nitrogen.

7.3.6.4. *Generalized chart for fugacity coefficients*

We can write relation [7.46] in the form:

$$\ln \varphi^0 = \frac{1}{\mathrm{R}T}\int_0^P \left(PV(P,T)-1\right)\frac{\mathrm{d}P}{P} \qquad [7.52]$$

But by introducing the coordinates of the critical point and the reduced coordinates (see relation 6.31), we obtain:

$$\frac{PV_m(P,T)}{\mathrm{R}T} = \frac{P_r V_r}{\mathrm{R}T_r}\frac{P_c V_c}{T_c} \qquad [7.53]$$

Equation [7.52] becomes:

$$\ln \varphi^0 = \int_0^{P_r}\left(\frac{P_r V_r}{\mathrm{R}T_r}\frac{P_c V_c}{T_c} - 1\right)\frac{\mathrm{d}P_r}{P_r} \qquad [7.54]$$

At a given reduced temperature, the second member of the equation is only a function of the reduced pressure, the same for the fugacity coefficient of any real gas considered. Therefore for each reduced temperature, constructing the curves $\phi^0 = F(P_r)$ and a single chart (Figure 7.5) allows us to determine the fugacity coefficient ϕ^0 if we know the coordinates of the critical point P_c and T_c. This is the generalized chart of fugacity coefficients (Figure 7.5).

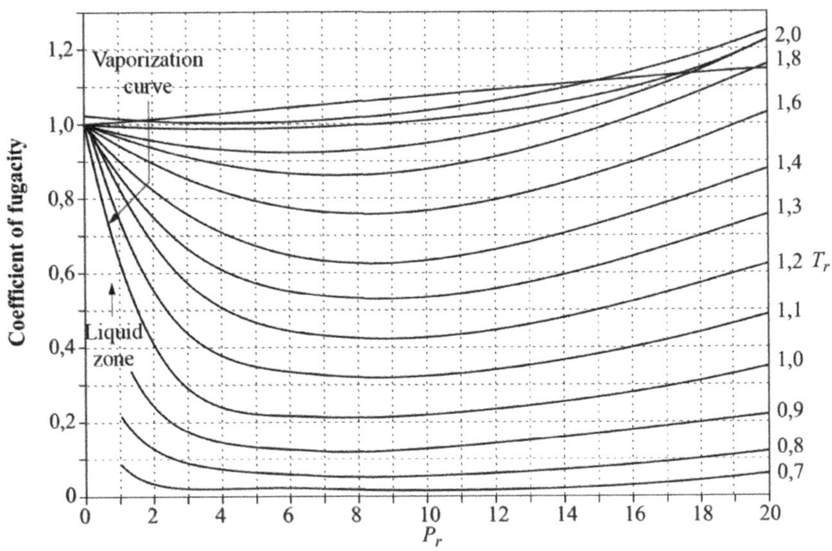

Figure 7.5. *Fugacity coefficient chart of pure gases*

So, for example, to determine the fugacity coefficient of a gas at 0°C at a pressure of 100 bars, we obtain the values for the critical temperature and pressure of this gas from the tables, we deduce the reduced temperature and reduced pressure. On the x-axis, at point P_r on the curve relative to T_r on the y-axis, φ^0 can be directly read from generalized charts of fugacity coefficients.

7.3.7. *Heat capacities of gases*

The previous data only provide three of four thermodynamic coefficients. We also need to determine the second diagonal coefficient which is $\left(\frac{\partial S}{\partial T}\right)_V$ or $\left(\frac{\partial S}{\partial T}\right)_P$, which means that we need to determine the molar heat capacities at a constant volume or pressure, defined by the relations:

$$C_V = \frac{1}{T}\left(\frac{\partial S}{\partial T}\right)_V \qquad [7.55]$$

and

$$C_P = \frac{1}{T}\left(\frac{\partial S}{\partial T}\right)_P \qquad [7.56]$$

7.3.7.1. *Heat capacities at constant pressure*

In practice, many tables provide the heat capacities of gases at constant pressure and atmospheric pressure in the form of a polynomial:

$$C_P = a + bT + cT^2 + dT^{-2} \qquad [7.57]$$

The coefficients a, b, c and d depend on the substance and theoretically depend on pressure. This dependence is not usually provided as it can be easily calculated (see section 7.2.7.3). Some coefficients of relation [7.57] can be zero and one of either coefficient c or d will always be zero.

7.3.7.2. *Heat capacities at constant volume*

The heat capacity at constant volume is deduced from the previous data at constant pressure since it involves moving from a coefficient of matrix $\mathrm{M}(P,T)$ to a coefficient of matrix $\mathrm{M}(V,T)$ (see section 2.1.5), we obtain:

$$C_P - C_V = T\left(\frac{\partial P}{\partial T}\right)_V \left(\frac{\partial V}{\partial T}\right)_P \qquad [7.58]$$

This difference can be calculated from the state equation or compressibility coefficient Z for example.

7.3.7.3. *Variations in heat capacities with pressure or volume*

From the symmetry of matrix $\mathrm{M}(P,T)$, it is easy to show that:

$$\frac{\partial C_P}{\partial P} = -T\left(\frac{\partial^2 V}{\partial T^2}\right)_P \qquad [7.59]$$

We also show that:

$$\frac{\partial C_V}{\partial V} = -T\left(\frac{\partial^2 P}{\partial T^2}\right)_V \qquad [7.60]$$

These two expressions are assessed by simply knowing the other thermodynamic coefficients, i.e. by the state equation or the compressibility coefficient Z for example.

7.4. Microscopic description of a real gas

Microscopically speaking, the gas is defined by its canonical partition function taking into account the terms linked to the internal energy, of translation and intermolecular interactions.

The microscopic description of a real gas considers a gas as a collection of identical quantum particles as part of the statistical approximation of the classical limiting case, i.e. as part of the approximation [7.61] which we have seen when establishing the partition function of molecules (see Chapter 5):

$$\exp(-\alpha) << 1 \qquad [7.61]$$

α is the Lagrange coefficient relative to the number of moles, an approximation which is most frequently used when establishing the molecular partition function. We will express the complete canonical partition function of the system to deduce the thermodynamic functions.

7.4.1. *Canonical partition function of a fluid*

The interaction energies between molecules is a complex issue and is most frequently satisfied by approximations. For example, we have seen that (see section 6.6) the interaction between two molecules does not disturb their internal organizations and the interactions forces only depend on the distance between the molecules (see section 6.10). Between two molecules, the corrective energetic term linked to the interaction ε_{AB} will appear, which is practically zero when both molecules A and B are far enough as in the example of the Lennard–Jones potential (see relations [6.100] or [6.103]). Calculating the partition function of a fluid, in particular the calculation of the second coefficient of the virial, is tricky; many examples in the literature are approximate or full of poorly justified shortcuts. This is why we chose to use the method developed by R. Fowler and E.A. Guggenheim and contributions by Ursell.

Let us consider a mixture of N independent molecules, without any interaction. This collection is treated as an assembly of non-discernible quantum particles in the classical limiting case of validity of relation [7.61].

If ω_i is the ensemble of positional coordinates $(x_1, x_2, x_3)_i$ of particle i, between two particles i and j there is an interaction energy $\varepsilon_{i,j}$ and we assume that the interaction energy only contains terms such as $\varepsilon_{i,j}$ even if several molecules are in mutual interaction. This hypothesis is known as the paired interaction model. The configuration integral (see relation [6.28]) has the following form:

$$I_I = \int_{Nv_f} \int \dots \int \prod_{i<j} \exp - \frac{\varepsilon_{i,j}}{k_B T} (d\omega)^N \qquad [7.62]$$

The integrals are extended to the volume available for molecules i.e. Nv_f.

We are going to introduce the Mayer function $y_{i,j}$, defined by:

$$y_{i,j} = \exp\left(-\frac{\varepsilon_{i,j}}{k_B T}\right) - 1 \qquad [7.63]$$

Function [7.62] becomes:

$$I_{(I)} = \frac{1}{N!} \int_{Nv_f} \int \dots \int \left(\prod_{j=1}^{N(N+1)/2} \left(\prod_{i=j+1}^{N(N+1)/2} (1 + y_{i,j}) \right) \right) (d\omega)^N \qquad [7.64]$$

The form of interaction functions such as those of Lennard–Jones, based on the model of Van der Waals forces involving Keesom orientation effects, Debye induction and London dispersion, which quickly decrease with distance; beyond a certain distance r_∞ between two molecules, the interaction can be negligible (for example, when the interaction is less than $5\varepsilon_0/100$). This comes down to defining around each molecule a volume influence:

$$v_\infty = \frac{\pi r_\infty^3}{6} \qquad [7.65]$$

To go beyond the exploitation of relation [1 .64], we are going to develop the product using the power series $y_{i,j}$, the configuration integral takes the following form:

$$I_{(I)} = \frac{1}{N!} \int_{Nv_f} \int \ldots \int \left(\begin{array}{c} 1 + \sum y_{i,j} + \sum y_{i,j} y_{i,k} \\ + \sum y_{i,j} y_{j,k} y_{k,i} + \sum y_{i,j} y_{k,l} + \ldots \end{array} \right) (\mathrm{d}\,\omega)^N \quad [7.66]$$

The different sums of the development correspond to different interactions:

– $\sum y_{i,j}$ groups the configurations in which a single interaction pair is involved, i.e. the case where a single molecule j is found within the sphere of influence of molecule i (Figure 7.6(a));

– $\sum y_{i,j} y_{i,k}$ groups the configurations in which two interaction pairs are involved, i.e. the case of two molecules j and k are found within the sphere of influence of the same molecule i without any of the molecules j or k being found within the sphere of influence of another (Figure 7.6(b));

– $\sum y_{i,j} y_{j,k} y_{k,i}$ groups the configurations in which three molecules are interacting in pairs (Figure 7.6(c));

– $\sum y_{i,j} y_{k,l}$ groups the configurations in which two different pairs of molecules interact via a single pair i, k (Figure 7.6(d));

– etc.

Figure 7.6. *Different configurations of paired molecules for the first four terms of the sum in relation [7.66]*

We are now going to constrain the sum of relation [7.66] only taking into consideration the interactions of one molecule with another molecule, which we call paired interactions excluding interactions between three and more molecules simultaneously, that is to say, considering Figures 7.6(a) and 7.6(b). This approximation is valid if Nv_0/V is small with regard to 1, i.e. if the gas is not too imperfect, the molecules are not too dense or if the pressure is not too high. We begin with a mixture of N independent molecules, without interaction. We introduce the existence of an interaction pair.

To materialize this paired interaction, we are going to use the quasi-chemical method (see Appendix 3) we are going to replace our system of N molecules of A with a mixture of N_1 molecules of A and N_2 molecules of dimer A_2 which are formed every time a molecule enters the spherical interaction volume with radius r of another molecule; let $\varepsilon(r)$ be the interaction energy between the two molecules of A in the dimer. We therefore assume that, according to the model of associated solutions, these two populations of monomers and dimers are in thermodynamic equilibria and form a perfect solution, i.e.:

$$N = N_1 + 2N_2 \qquad [7.67]$$

and the application of the equilibrium condition gives:

$$\frac{N_2}{N_1^2} = K_c \qquad [7.68]$$

Let us express the equilibrium constant of the association reaction using the partition functions of reactants and products in the following form using relation [A.3.10] with $E_0 = 0$ since the origins of energies of species A and A_2 are the same:

$$\frac{N_2}{N_1^2} = \frac{z_2(T)}{[z_1(T)]^2} \qquad [7.69]$$

The partition functions z_1 and z_2 can be expressed using the overall partition function of molecule A.

With regard to z_1, in the partition function of a real monomer, the volume V is replaced by volume V_1 available to molecules in our mixture; this is in fact the reduced volume V of the volume of spheres with a radius of influence r_0 between two molecules and V_1 will therefore be:

$$V_1 = V - \frac{4\pi r_0^3 N}{6} \qquad [7.70]$$

Coefficient 6 (2x3) of the denominator is derived from counting a molecule of A, once as neighboring and once as in the center of the sphere of influence.

The partition function of molecules of the monomer will therefore be, by separating the translational contribution (z_t) of the rest of the partition function ($z_{int}(T)$):

$$z_1 = \frac{z_{int}(T)(z_t)^3}{V} V_1 \tag{7.71}$$

The volume available for dimers is the total volume V. If we account for the interaction, and if we use relation [6.5] for the corresponding term, which is the potential energy, the partition function of a dimer molecule will be:

$$z_2 = [z_{\text{int}}(T)]^2 (z_t)^3 \iiint_V \exp\left(-\frac{\varepsilon(r)}{k_B T}\right) \mathrm{d}x.\mathrm{d}y.\mathrm{d}z \tag{7.72}$$

x, y and z are the relative coordinates of a molecule A with regard to that at the center of the sphere of integration. With regard to polar coordinates, relation [7.72] becomes:

$$z_2 = \frac{[z_{\text{int}}(T)]^2 (z_t)^3}{2} \int_0^{r_0} 4\pi r^2 \exp\left(-\frac{\varepsilon(r)}{k_B T}\right) \mathrm{d}r \tag{7.73}$$

Coefficient 2 of the denominator is derived from a number of symmetry in the partition function from the identity of two A molecules composing the dimer.

In a simpler form, we put A(T) as:

$$A(T) = \frac{1}{2} \int_0^{r_0} 4\pi r^2 \left[\exp\left(-\frac{\varepsilon_{\text{AA}}}{\text{k}_\text{B} T}\right) - 1\right] \mathrm{d}r \tag{7.74}$$

By integrating in parts, we obtain:

$$A(T)=\frac{2\pi r_0^3}{3}-\frac{1}{2}\int_0^\infty 4\pi r^2\left[\exp\left(-\frac{\varepsilon_{AA}}{k_B T}\right)-1\right]dr=\frac{2\pi r_0^3}{3}+B_{AA}(T) \qquad [7.75]$$

Let:

$$B_{AA}(T)=-\frac{1}{2}\int_0^\infty 4\pi r^2\left[\exp\left(-\frac{\varepsilon_{AA}}{k_B T}\right)-1\right]dr \qquad [7.76]$$

By taking into account expressions [7.66], [7.68], [7.69], [7.71], [7.72], [7.73] and [7.74], it becomes:

$$N_1=N-\frac{2N^2 A(T)}{V} \qquad [7.77]$$

and

$$N_2=\frac{N^2 A(T)}{V} \qquad [7.78]$$

The canonical partition function of our ensemble (non-localizable), taking into account the presence of two non-interacting components, will be:

$$Z_C=\frac{(z_1)^{N_1}}{N_1!}\frac{(z_2)^{N_2}}{N_2!} \qquad [7.79]$$

By applying Stirling's approximation it becomes:

$$\ln Z_C=N_1\left[\ln\frac{z_1}{N_1}+1\right]+N_2\left[\ln\frac{z_2}{N_2}+1\right] \qquad [7.80]$$

which leads to, by revealing the amount n of gas:

$$\ln Z_C=n\mathrm{N_a}\ln\left(\frac{z_{pf}}{n\mathrm{N_a}}+1\right)-\frac{n^2\mathrm{N_a}^2 B_{AA}(T)}{V} \qquad [7.81]$$

Without any interaction and in the hypothesis of quantum statistics in the classical limiting case, relation [7.75] gives the canonical partition function, which we can write for a perfect pure gas in the form:

$$\ln Z_{C(pf)} = n\mathrm{N_a} \ln z_{pf} - n\mathrm{N_a} \ln\lfloor (n\mathrm{N_a})! \rfloor \quad [7.82]$$

Using Stirling's formula [4.1]:

$$\ln n! \cong n(\ln n - 1) \cong n \ln n \quad [7.83]$$

This relation becomes:

$$\ln Z_{C(pf)} = n\mathrm{N_a} \ln\left(\frac{z_{pf}}{n\mathrm{N_a}} - 1 \right) \quad [7.84]$$

The partition function takes into account interactions between atoms of a fluid, i.e. the exchange integral is written by the difference between [7.81] and [7.84]:

$$\ln I_{C(\mathrm{AA})} = -\frac{n^2 \mathrm{N_a}^2 B_{\mathrm{AA}}(T)}{V} \quad [7.85]$$

Note that the lower limit of the integration of equation [7.76] is 0 when the distance between molecules tends towards zero, all known expressions of potential, particularly those of Lennard–Jones, leads to an integral that diverges. To overcome this issue, we will only count the potentials from distance *D*. This distance forms an impenetrable wall to a neighboring molecule. The volume of the corresponding sphere will be considered as an incompressible volume of the molecule, i.e.:

$$v_0 = \frac{\pi D^3}{6} \quad [7.86]$$

The free volume available for molecules becomes:

$$N v_f = V - N v_0 \quad [7.87]$$

and the interaction term becomes:

$$\ln I_{(AA)} = -\frac{N^2 B_{AA}(T)}{Nv_f} = \frac{n^2 \mathrm{N_a}^2 B_{AA}(T)}{Nv_f} \qquad [7.88]$$

i.e. by taking into account [7.87]:

$$\ln I_{(AA)} = -\frac{n^2 \mathrm{N_a}^2 B_{\mathrm{AA}}(T)}{V - Nv_0} \qquad [7.89]$$

To calculate $B_{\mathrm{AA}}(T)$, we must know the laws of energy $\varepsilon_{\mathrm{AA}}$ variations with distances between molecules.

7.4.2. *Helmholtz energy and development of the virial*

We know that the Helmholtz energy (relation [5.43]) is expressed using the following canonical partition function:

$$F = U - TS = -\frac{\ln Z_C}{\beta} = -\mathrm{k_B} T \ln Z_C \qquad [7.90]$$

By combining this relation and relation [7.89], the Helmholtz energy of the perfect gas is written:

$$F = -n\mathrm{R}T\left(\ln\frac{z_{pf}}{n\mathrm{N_a}} + 1\right) - \frac{\mathrm{R}T}{V - Nv_0} n^2 \mathrm{N_a} B_{\mathrm{AA}}(T) \qquad [7.91]$$

The first term of this expression is in fact the Helmholtz energy of the perfect gas.

According to the definition of the Helmholtz energy function, we can easily deduce that the pressure is given by the opposite of the derivative of this Helmholtz energy with regard to volume:

$$P = -\frac{\partial F}{\partial V} \qquad [7.92]$$

The term z_{pf} of equation [7.91], which represents the partition function of the non-interacting gas molecule, contains a volume resulting from the contribution of three degrees of translation to the partition function; we therefore write the Helmholtz energy in the form:

$$F = -n\mathrm{R}T\left(\ln\left(\frac{z_{pf}}{V}\right)\frac{V-2Nv_0}{n\mathrm{N_a}}+1\right)-\frac{\mathrm{R}T}{V-Nv_0}n^2\mathrm{N_a}B_{\mathrm{AA}}(T) \quad [7.93]$$

Knowing that the term z_{pf}/V is independent of volume and that volume $V-Nv_0$ is small compared to V, we can write the approximations:

$$(V-Nv_0)^2 \approx V(V-2Nv_0) \text{ and } (V-Nv_0)\approx V$$

By applying equation [7.92], we obtain:

$$P(V-2Nv_0) = n\mathrm{R}T\left(1+\frac{n\mathrm{N_a}B_{\mathrm{AA}}(T)}{V}\right) \quad [7.94]$$

We can see that this expression is identical to the virial equation given by relation [7.20], limited to the second term; by identification, the second coefficient of the virial is therefore:

$$B_2(T) = \mathrm{N_a}B_{\mathrm{AA}}(T) \quad [7.95]$$

The term B_{AA}(T) represents the second coefficient of the virial reduced to one molecule.

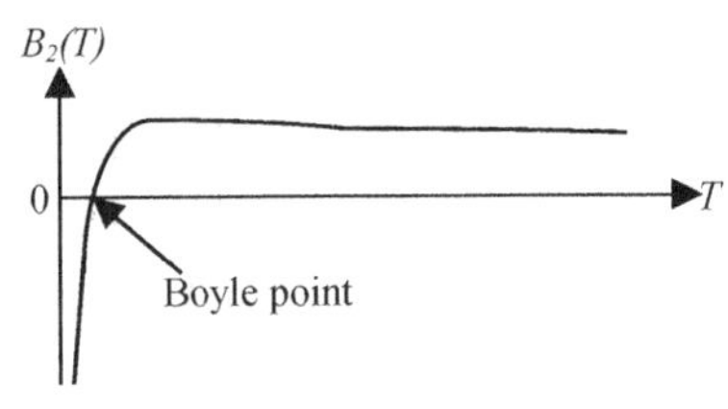

Figure 7.7. *Variation in the second coefficient of the virial with temperature*

Our real gas model, taking into account only the paired interactions of central forces between molecules, limits our representation of a real gas to the second coefficient of the virial. So, the different energetic terms with two, three and four components, which we introduced in section 7.3.1, correspond to the second, third, fourth etc. coefficient of the virial. For the coefficients following the second coefficient, the calculations quickly become inextricable and no longer give analytical expressions and only a molecular dynamic approach can go beyond this, but it does not give a physical meaning to the other coefficients of the virial, which become empirical coefficients of the development in series of a function that cannot be expressed in an analytical way. However, we will express the second coefficient according to temperature using the forces between molecules to better our physical understanding of a real gas, even if this representation will still be very imperfect. Figure 7.7 shows the variation in the second coefficient of the virial according to temperature. We find that the coefficient is cancelled out at a point called the Boyle point, and the curve has a maximum.

7.4.3. *Forms of the second coefficient of the virial*

To explain the second coefficient of the virial, we must express the function $\varepsilon(r)$. To do this, we will gradually review three models of interactions between molecules, which ultimately lead to successive developed models of this coefficient.

7.4.3.1. *Hard sphere model without interaction force*

In this simple model, each molecule is supposedly protected by a sphere with diameter D and another molecule cannot cross this sphere (Figure 7.8(a)), this results in a function $\varepsilon(r)$ similar to that in Figure 7.8(b), which is defined by:

$$\begin{cases} \varepsilon(r): \textbf{infinite} \quad \textbf{if}\, r < D \\ \varepsilon(r) = 0 \quad \textbf{if}\, r \geq D \end{cases} \qquad [7.96]$$

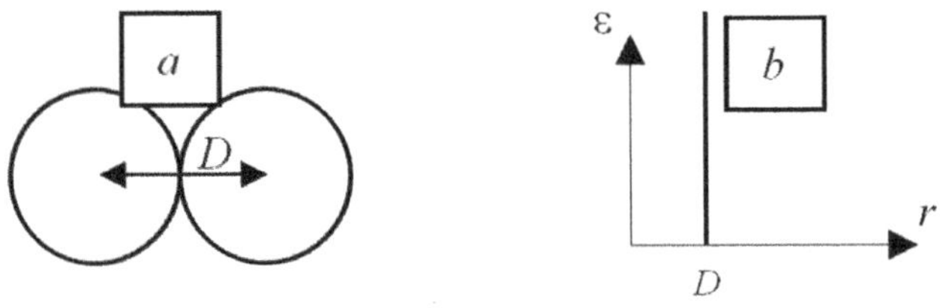

Figure 7.8. *Hard-sphere model without attraction*

Equation [7.76], which can be split into two parts, is written:

$$B_{AA}(T) = \frac{1}{2}\int_0^D 4\pi r^2 \left[1 - \exp\left(-\frac{\varepsilon}{RT}\right)\right] \mathrm{d}r + \frac{1}{2}\int_D^\infty 4\pi r^2 \left[1 - \exp\left(-\frac{\varepsilon}{RT}\right)\right] \mathrm{d}r \quad [7.97]$$

So, by calculating the two integrals:

$$B_{AA}(T) = \frac{1}{2}\frac{4\pi(D)^3}{3} + 0 = \frac{2\pi D^3}{3} \quad [7.98]$$

In this rather simplified model, the second coefficient of the virial is a constant and does not therefore depend on temperature.

7.4.3.2. *Keesom hard-sphere model without repulsion force*

In this model, out of the hard sphere of protection, there is an attraction energy of $1/r^6$, function $\varepsilon(r)$ is represented on Figure 7.9 and is defined by, if ε_0 is the negative value of the mutual energy of both protection spheres of two molecules in contact:

$$\begin{cases} \varepsilon(r): \textbf{infinite} & \textbf{if } r < D \\ \varepsilon(r) = \varepsilon_0 \left(\dfrac{D}{r}\right)^6 & \textbf{if } r \geq D \end{cases} \quad [7.99]$$

This definition gives a molecular diameter D which acts as the contact diameter of contact molecules.

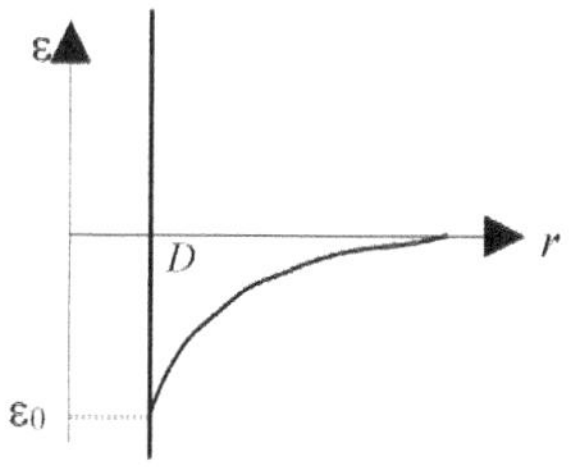

Figure 7.9. *Hard-sphere model with force of attraction*

To integrate expression [7.76], we develop the exponential in series:

$$\exp\left(-\frac{\left(\frac{D}{r}\right)^6}{\mathrm{k_B}T}\right)=\sum_{i=0}^{\infty}\frac{1}{i!}\left(-\frac{\varepsilon_0 D^6}{\mathrm{k_B}Tr^6}\right)^i \qquad [7.100]$$

Equation [7.72] is then split into two parts and the integration gives:

$$B_{\mathrm{AA}}(T)=\frac{2\pi D^3}{3}\left\{1-\sum_{i=1}^{\infty}\frac{1}{(2i-1)i!}\left(-\frac{\varepsilon_0}{\mathrm{k_B}T}\right)^i\right\} \qquad [7.101]$$

This is the Keesom equation.

To simplify the expressions, we can keep the first term of the equation ($\varepsilon_0 << \mathrm{k_B}T$), which gives:

$$B_{\mathrm{AA}}(T)=\frac{2\pi D^3}{3}\left\{1+\frac{\varepsilon_0}{\mathrm{k_B}T}\right\} \qquad [7.102]$$

In the Keesom model, the second coefficient of the virial is approximately a linear equation of the inverse of temperature.

7.4.3.3. *The Lennard-Jones model*

The second coefficient of the virial was calculated by Lennard-Jones using, in relation [7.76], the potential function deduced from the van der Waals forces model, i.e.:

$$\frac{\varepsilon}{\varepsilon_0} = -2\left(\frac{r_0}{r}\right)^6 + \left(\frac{r_0}{r}\right)^{12} \qquad [7.103]$$

It revealed that this coefficient takes the form:

$$B_{\mathrm{AA}}(T) = \frac{\frac{2}{\sqrt{2}}\pi r_0^3}{3} J(y) = \frac{2\pi (D)^3}{3} J(y) \qquad [7.104]$$

The variable y is defined by:

$$y = -2\left(\frac{\varepsilon_0}{\mathrm{k_B}T}\right)^{1/2} \qquad [7.105]$$

Whereas function $J(y)$ obeys relation:

$$J(y) = y^{1/2}\left\{\Gamma(3/4) - \frac{1}{4}\sum_{i=1}^{\infty} \Gamma\left(\frac{2i-1}{4}\right)\frac{y^i}{i!}\right\} \qquad [7.106]$$

In this expression $\Gamma(x)$ is the value of function Γ for a value x of the variable (see Appendix A.5), for example, we obtain $\Gamma(3/4) = 6.9$.

From [7.104] and [7.106], we can write:

$$\log\left|B_{\mathrm{AA}}(T)\right| - \log\left|J(y)\right| = \log\frac{2\pi}{3} + 3\log D \qquad [7.107]$$

And from [7.105], we obtain:

$$\log T - (-2\log y) = -\log \varepsilon_0 + \log(\mathrm{k_B}/4) \qquad [7.108]$$

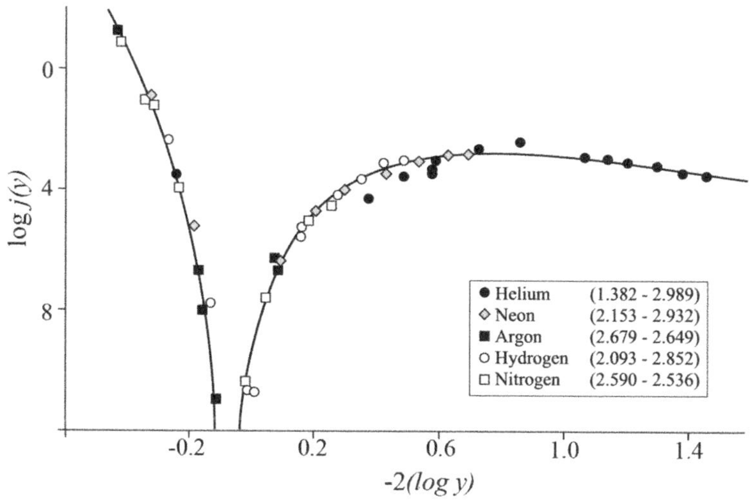

Figure 7.10. *Variations in the second coefficient of the virial with temperature (according to Lennard-Jones)*

Both expressions [7.105] and [7.108] show that the representative curves of functions $\log(|J(y)|) = f(-2\log y)$ and $\log|B_{AA}(T)| = g(T)$ can be superimposed. This superposition allows us to determine parameters ε_0 and D and therefore r_0 depends on the gas. Figure 7.10 shows that for several gases, the law is followed correctly.

For example, for nitrogen we determine:

$$\frac{|\varepsilon_0|}{k_B} = 96.7 \ , \ D = 3.72 \text{ Ä} \ \text{ and } \ r_0 = 4.17\text{Ä}$$

We can easily determine the Boyle point and obtain:

$$y_{Bo} = 1.082$$

and by applying relation [7.105] we obtain:

$$\frac{k_B T_{Bo}}{|\varepsilon_0|} = 3.43$$

Combining relations [7.104], [7.105] and [7.106], for the second coefficient of the virial, gives:

$$B_{\mathrm{AA}}(T)=-\frac{2\pi r_0^3}{3}\left(\frac{\varepsilon_0}{\mathrm{k_B}T}\right)^{1/4}\left\{6.9+\frac{1}{4}\sum_{i=1}^{\infty}\frac{2^i}{i!}\left(\frac{\varepsilon_0}{\mathrm{k_B}T}\right)^{i/2}\Gamma\left(\frac{2i-1}{4}\right)\right\} \qquad [7.109]$$

NOTE.– If we limit the sum of the previous relation to its first term, we have a simplified form of the second coefficient:

$$B_{\mathrm{AA}}(T)=-\frac{13.8\pi r_0^3}{3}\left(\frac{\varepsilon_0}{\mathrm{k_B}T}\right)^{1/4}\left\{1+0.87\left(\frac{\varepsilon_0}{\mathrm{k_B}T}\right)^{1/2}\right\} \qquad [7.110]$$

7.4.4. *Macroscopic state equations and microscopic description*

We are now going to focus on the second coefficient of the virial, calculated from the microscopic properties of macroscopic state equations, to ultimately give a physical meaning to the parameters of these state equations. We are therefore going to assess the application of relations [7.97], [7.102] and [7.110] to relation [7.96] to calculate the second coefficient, then to [7.94] to compare several state equations.

If we use relation [7.98], we obtain the following state equation:

$$P(V-2Nv_0)=n\mathrm{R}T\left(1+\frac{2\pi \mathrm{N_a} D^3}{3v^0}\right) \qquad [7.111]$$

which is equivalent to:

$$P(v^0-b)=\mathrm{R}T \qquad [7.112]$$

which is the simplified Van der Waals equation (with $a = 0$) and therefore the co-volume is, by taking into account [7.21]:

$$b=\frac{\pi \mathrm{N_a} D^3}{3} \qquad [7.113]$$

Let us now assess the use of the simplified Keesom equation [7.102]:

$$Pv^0 = \mathrm{R}T\left(1+\frac{\pi N_a D^3}{3v^0}+\frac{\pi\varepsilon_0 N_a D^3}{3\mathrm{k_B}Tv^0}\right) \qquad [7.114]$$

If we compare this to the Van der Waals equation in the form of a development of the virial (relation [7.21]), we obtain:

$$b=\frac{\pi N_a D^3}{3} \text{ and } a=\frac{\pi\varepsilon_0 \mathrm{N_a^2} D^3}{3}=\mathrm{N_a} b\varepsilon_0 \qquad [7.115]$$

So, we see that the Van der Waals equation contains at least the assumption $\varepsilon_0 << \mathrm{k_B}T$.

The form [7.19] of the Dieterici equation gives the same meaning as the van der Waals to the second coefficient of the virial.

Now let us assess the use of the simplified Lennard–Jones equation (equation [7.110]), if we keep the same meaning [7.115] for the co-volume, we obtain:

$$Pv^0 = \mathrm{R}T\left[1-\frac{bNa}{v^0}\left(\frac{6.9\varepsilon_0^{1/4}}{\mathrm{k_B^{1/4}}T^{1/4}}+\frac{1.86\varepsilon_0^{3/4}}{\mathrm{k_B^{3/4}}T^{3/4}}\right)\right] \qquad [7.116]$$

We obtain a state equating with the unrecorded form in the macroscopic functions but more complex than that of Berthelot.

7.4.5. *Chemical potential and fugacity of a real gas*

Using equation [7.93], we calculate the chemical potential (or molar Gibbs energy) using the following relation:

$$\mu=\left(\frac{\partial F}{\partial n}\right)_{T,V}=\mathrm{R}T\ln\frac{n\mathrm{N_a}}{z_{pf}}-\frac{2\mathrm{R}T}{V}n\mathrm{N_a}B_{\mathrm{AA}}(T) \qquad [7.117]$$

We see in the first term in the second half of equation [7.117] the chemical potential of the perfect gas. By comparing this to the definition of relation [7.43], by taking into account [7.117], we can write:

$$f_0 = \frac{n\mathrm{R}T}{V} - \frac{2\mathrm{R}T}{V}\exp\frac{2n\mathrm{N_a}B_{\mathrm{AA}}(T)}{V} \qquad [7.118]$$

This expression gives the fugacity of the pure gas as a function of the second coefficient of the virial.

Lewis created an approximate formula to calculate coefficient of fugacity, by developing relation [7.118] in series and by being limited to the first term, we obtain:

$$\varphi_0 = \frac{f_0}{P} \cong 1 + \frac{n\mathrm{N_a}B_{\mathrm{AA}}(T)}{V} \qquad [7.119]$$

So, let P^* be the pressure of a gas with perfect behavior, we obtain:

$$\frac{P}{P^*} \cong 1 + \frac{n\mathrm{N_a}B_{\mathrm{AA}}(T)}{V} \qquad [7.120]$$

So, the Lewis equation is:

$$\varphi_0 = \frac{f_0}{P} \cong \frac{P}{P^*} \qquad [7.121]$$

We can also easily show that:

$$f_0 - P \cong P - P^* \qquad [7.122]$$

7.4.6. *Conclusion on microscopic modeling of a real gas*

We are limited, when determining the second coefficient of the virial, to spherical molecules. Although this model is acceptable for simple gases or polyatomic gases with an almost spherical shape such as ammonia, other models should be developed for molecules with elongated forms and for polymers.

With regard to the imperfection of gases, we are limited to the forces between two molecules only (two body forces) which give the expressions of the second coefficient of the virial. Researchers have endeavored to calculate the third, fourth and fifth coefficient of the virial. Here the three body forces are involved for the third coefficient, four body forces for the fourth and compact packing of spheres for the fifth coefficient. As for the second coefficient, the authors initially stuck to the hard-sphere model without attraction force (see section 7.3.3.1 and Figure 7.6), and as in the case of the second coefficient, they obtained coefficients practically independent of temperature, which allowed Hirshfelder and Roseveare to propose a state equation in the form:

$$PV = RT\left\{1+\frac{B_2}{v^0}+\frac{0.625b^2}{\left(v^0\right)^2}+\frac{0.2869b^3}{\left(v^0\right)^3}+\frac{0.1928b^4}{\left(v^0\right)^4}\right\} \qquad [7.123]$$

In this expression, B_2 is the second coefficient calculated as we saw above and b is the co-volume, which is involved in the Van der Waals equation (equation [7.113]). As we initially did to calculate the second coefficient, the authors calculated the following coefficients using the expression of two Lennard–Jones constants for the interaction between molecules:

$$\varepsilon = -\frac{m_0}{r^6}+\frac{l_0}{r^{12}} \qquad [7.124]$$

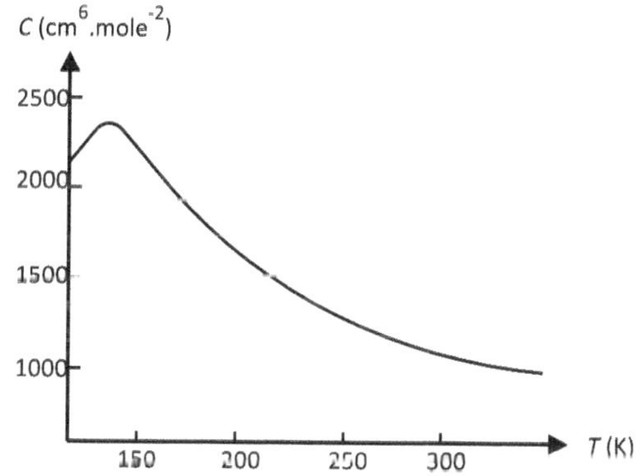

Figure 7.11. *Variation in the third coefficient of the virial with temperature according to Malijevsky et al.*

For variations in the third coefficient with temperature, Malijevsky and his team arrive at the curve in Figure 7.11, which essentially shows a decrease in this coefficient with temperature. According to Hall, the fourth and fifth coefficient will be proportional to the different powers of the second coefficient:

$$B_4 = 0.02820(B_2)^2 \text{ and } B_5 = -0.0104(B_2)^3 \qquad [7.125]$$

Ultimately, the coefficients higher than the third may have the potential of being used for liquids, but with regard to gases, the experimental properties are correctly described by the development of the virial limited to the third coefficient. Not that this is not the case for the Van der Waals equation.

7.5. Microscopic approach of the heat capacity of gases

To completely model the heat capacity of a real gas, the heat capacity must be modeled at constant volume for a perfect gas; in fact, all gases tend towards the perfect state when the pressure tends towards zero or the volume tends towards infinity. We will establish the law giving $C_V(T)$ for the perfect gas, then using relation [7.58] and integrating for volumes less than V, we obtain, averaging the correct representation of thermomechanical properties, the heat capacity at constant volume for a gas at any volume. In addition, we move from $C_V(T)$ to $C_P(T)$ using relation [7.56] and integrating the inequalities [7.57] between the pressures 0 and P. We are going to therefore model the heat capacity of a perfect gas, for which we will use the relation of this capacity from the internal energy:

$$C_V = \left(\frac{\partial U}{\partial T}\right)_{N,V} \qquad [7.126]$$

As the internal energy of a perfect gas is the sum of the contributions of different modes (translation, rotation, vibration), we can write:

$$C_V = C_{V(t)} + C_{V(r)} + C_{V(v)} \qquad [7.127]$$

And for each mode m, the contribution will be:

$$C_{V(m)} = N_a \left[\frac{\partial \langle \varepsilon_m \rangle}{\partial T}\right] \qquad [7.128]$$

$\langle \varepsilon_m \rangle$ is the mean energy corresponding to the contribution m.

7.5.1. *Classical theorem from the equipartition of energy*

The first values for the heat capacities have been calculated, by taking into account for each mode, the mean energy corresponding to the equipartition of energy (see section 6.8). Using the values given, respectively, by equations [6.71] for the translation, [6.74] for the rotation and [6.78] for the vibration, we obtain for a diatomic molecule:

$$C_V = \frac{3}{2}R + R + R = \frac{7}{2}R \qquad [7.129]$$

Molecule	v	r	$C_{V(t)}$	$C_{V(r)}$	$C_{V(v)}$	C_V
Monoatomic	0	0	3R/2	0	0	3R/2
Diatomic	1	1	3R/2	R	R	7R/2
Linear triatomic	5	1	3R/2	R	5R	15R/2
Non-linear triatomic	4	2	3R/2	2R	4R	15R/2
Non-linear tetratomic	7	2	3R/2	2R	7R	21R/2
Non-linear pentatomic	8	2	3R/2	2R	8R	23R/2

Table 7.2. *Heat capacities at volume constant in the classical theorem of equipartition of energy*

For a more complex molecule, we multiply the values of each mode by the number of degrees of freedom corresponding to this mode; so, a linear molecule will have only one degree of rotational freedom whereas a non-linear molecule will have two. As the number of degrees of translational freedom is always 3, the number of degrees of vibrational freedom of a molecule with N atoms is given by:

$$v = 3N - r - 3 \qquad [7.130]$$

Table 7.2 gives the values calculated for different types of molecules.

Of course, we can easily deduce the corresponding values for the heat capacities at constant pressure by adding the value R to the values in Table 7.2.

The experiment shows that the calculated values are generally too high with regard to the true values of the heat capacities. However, also note that if we disregard the vibrational term then the calculated values are this time too low. In fact, we obtain values close to the measurements at higher temperatures when there is no dissociation of the molecule.

7.5.2. *Quantum theorem of heat capacity at constant volume*

We are going to use relation [7.127] but by introducing the mean values of the different energies obtained at lower temperatures.

The temperature is always high enough so that the mean translational energy keeps the value given by relation:

$$\langle \varepsilon_t \rangle = \frac{3}{2} \mathrm{k_B} T \qquad [7.131]$$

With regard to the rotational term, if the temperature is lower than the characteristic temperature of rotation, this mode is not excited and its contribution is zero. For higher temperatures, we must use expression [6.51]:

$$z_r = \frac{T}{\Theta_r} = \frac{8\pi^2 I \mathrm{k_B} T}{\mathrm{h}^2} \qquad [7.132]$$

and calculate the mean energy using relation [6.69]:

$$\langle \varepsilon_m \rangle = -\frac{1}{z_m}\left(\frac{\partial z_m}{\partial \beta}\right) = \frac{\mathrm{k_B} T^2}{z_m}\left(\frac{\partial z_m}{\partial T}\right) \qquad [7.133]$$

We obtain a rotational heat capacity that varies with temperature as shown in the curve in Figure 7.12(a). For temperatures greater than the characteristic rotational temperature we find a contribution of $\mathrm{k_B}T$.

With regard to the vibrational term, the mean energy can be expressed by expression [6.76]:

$$\langle \varepsilon_v \rangle = \frac{\mathrm{h}\nu}{\exp\left(\frac{\mathrm{h}\nu}{\mathrm{k_B} T}\right) - 1} \qquad [7.134]$$

For a diatomic molecule this gives the expression of the heat capacity at constant volume given by:

$$C_{V(v)} = R\frac{\Theta_E}{T}\left\{\frac{\exp(-\Theta_E/2T)}{1-\exp(-\Theta_E/T)}\right\} \quad [7.135]$$

The curve in Figure 7.12(b) gives the variations of this contribution with temperature. If it is below the Einstein temperature, defined by:

$$\Theta_E = \frac{h\nu}{k_B} \quad [7.136]$$

Then the vibrational mode is not excited and the contribution of vibration is zero. At very high temperatures, we find the value of the equipartition of energy.

If we sum the three contributions to calculate the value for the heat capacity at constant volume, as the characteristic temperatures of rotation are often lower than the Einstein temperature (see Table 7.3), the variation in the heat capacity, for example of a diatomic molecule, with temperature takes the form of the curve in Figure 7.12(c). At low temperatures, the only contribution is that of translation, given by 3R/2. Then if the temperature increases, the contribution of rotation, is added according to the curve in Figure 7.12(a) until the limiting value of this contribution is reached, then the vibrational contribution is involved until the molecule dissociates which makes the heat capacity become double that of the translational contribution of monoatomic molecules. The limiting value of the vibrational contribution is sometimes never reached. This explains why the values calculated in Table 7.2 are too low if we do not take into account the vibration and too high in the opposite case.

Molecule	H_2	HCl	N_2	O_2	Cl_2
Θ_E(°K)	6.130	4.200	3.350	2.224	801
Θ_r(°K)	85	14	2.8	2.0	0.35

Table 7.3. *Values of Einstein temperature and characteristic temperature rotation for several molecules*

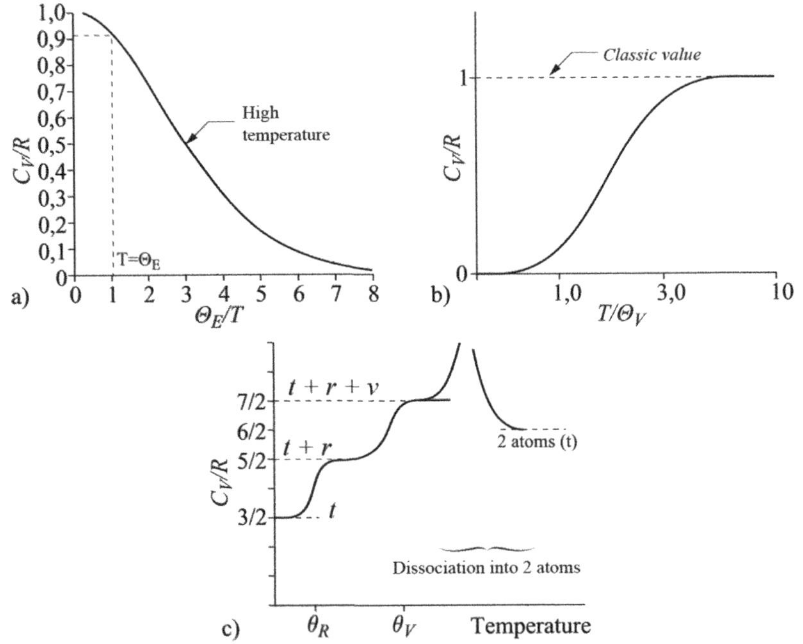

Figure 7.12. *Variation in heat capacity: (a) rotational contribution; (b) vibrational contribution; (c) total capacity*

In the case of molecules with more than two atoms, we know that the vibrational partition function is the product of the partitions functions relative to each mode of vibration and given by relation (relation [6.44]):

$$z_v \cong \prod_E \frac{\exp\left(-\dfrac{\Theta_E}{2T}\right)}{1-\exp\left(-\dfrac{\Theta_E}{T}\right)} \qquad [7.137]$$

Therefore, it is necessary to add as many vibrational contributions as there are degrees of vibrational freedom. Each term is characterized by its Einstein temperature, whereby the vibrational term is written:

$$C_{V(v)} = \mathrm{R}\sum_E \frac{\Theta_E}{T}\left\{\frac{\exp(-\Theta_E/2T)}{1-\exp(-\Theta_E/T)}\right\} \qquad [7.138]$$

The values obtained from the quantum theory are very close to the experimental values, even if some corrective terms are often included as a contribution due to hindered rotations, the anharmonicity of vibrations and changes in electron state requiring the electronic contribution to the heat capacity.

8

Gas Mixtures

Although a mixture of gases is only a special type of solution, it is often treated separately given the strong impact of pressure on its properties. Nevertheless, we will see, at the end of this chapter, that modern models of solutions, in particular the UNIFAC model, are also used when modeling gas mixtures.

8.1. Macroscopic modeling of gas mixtures

The general laws of solutions are applicable to gas mixtures, but there may be the possibility of generating more accurate models for the influences of pressure, which would provide new forms for the chemical potential.

8.1.1. *Perfect solutions of perfect gases*

The concept of perfection is not the same for a gas as it is for a solution; we must therefore be incredibly precise when using this term in the case of gas mixtures.

8.1.1.1. *Thermodynamic definition*

A gas mixture is said to be perfect if the chemical potential of each component i satisfies the perfect solution, which is written, as a function of its molar fraction y_i in the mixture:

$$\mu_i = g_i^0(T,P) + \mathrm{R}T \ln y_i \qquad [8.1]$$

If each gas is perfect, then the molar Gibbs energy of each gas in the pure state is given by:

$$g_i^0(T,P) = g_i^*(T,P^0) + RT\ln\frac{P}{P^0} \qquad [8.2]$$

$g_i^*(T,P^0)$ is the molar Gibbs energy of the pure gas i under a pressure of 1 bar; this is a function of temperature only.

By combining relations [8.1] and [8.2] it becomes:

$$\mu_i = g_i^*(T) + RT\ln\frac{P}{P^0} + RT\ln y_i \qquad [8.3]$$

By defining the partial pressure by relation:

$$P_i = Py_i \qquad [8.4]$$

we obtain:

$$P = \sum_{i=1}^{N} P_i \qquad [8.5]$$

By combining expressions [8.3] and [8.4], the chemical potential of component i is written:

$$\mu_i = g_i^*(T,P^0) + RT\ln\frac{P_i}{P^0} \qquad [8.6]$$

8.1.1.2. *Property*

Of course, these mixtures have all the properties of perfect solutions, but some of them can be explained according to partial pressures. Thus, the molar volume of the mixture obeys the ideal gas law, and we can write:

$$\left(\frac{\partial \mu_i}{\partial P}\right) = \overline{V_i} = v_i^0 \text{ or } V = \sum_{i=1}^{N} n_i\overline{V_i} \text{ whereby } V = \frac{RT}{P}\sum_{i=1}^{N} n_i = \frac{nRT}{P} \qquad [8.7]$$

v_i^0 is the molar volume of the pure gas i.

Similarly, we obtain:

$$P_i = Py_i = y_i \frac{n\mathrm{R}T}{V} = \frac{n_i \mathrm{R}T}{V} \qquad [8.8]$$

The gas i behaves as if it was alone at pressure P_i in the total volume V.

8.1.2. *Mixture of real gases*

A mixture of gases can be characterized by an equation of state in the form:

$$V = V(T, P, n_1, n_2, ..., n_i, ..n_N) \qquad [8.9]$$

The partial molar volume of a component can be expressed using this equation of state by deriving function [8.9] according to:

$$\overline{V_i} = \left(\frac{\partial V}{\partial n_i}\right)_{T,P,n_{j \neq i}} \qquad [8.10]$$

As we did for pure gases (see section 7.3.5), we determine the Gibbs energy of the mixture in the same way, obtaining a similar relation to equation [7.41]:

$$G(T,P,n_1...n_i...n_N) = G^*(T,P,n_1...n_i...n_N) + \lim_{P^0 \to 0} \int_{P^0}^{P} \left(V - V^*\right) \mathrm{d}P \qquad [8.11]$$

The asterisk indicates that the corresponding variable is relative to the perfect gas. Therefore, knowing the Gibbs energy of the mixture of perfect gases $G^*(T, P, n_1..n_i..n_N)$ and the state equation of the real mixture $V = V(T, P, n_1, n_2,..., n_i,...n_N)$, we can calculate the Gibbs energy of this mixture. By deriving relation [6.11], with regard to the amount of matter of one component i of the mixture, we obtain the chemical potential of this component:

$$\mu_i = g_i^* + \lim_{P^0 \to 0} \int_{P^0}^{P} \left(\overline{V_i} - \overline{V_i^*} \right) \mathrm{d}P \text{ with } \overline{V_i^*} = v_i^0 = \frac{\mathrm{R}T}{P} \qquad [8.12]$$

By introducing the fugacity f_i of the gas i in the mixture, analogous to the expression of the chemical potential of a perfect pure gas, we can write that of a gas i of a real mixture in convention (I), pure reference compound, in the form:

$$\mu_i(T,P) = g_i^*(T,P^0) + \mathrm{R}T \ln \frac{f_i(T,P,n_1,\ldots,n_i,\ldots n_N)}{P^0} \qquad [8.13]$$

Combining relation [8.13] and the corresponding relation for the pure component i, we obtain:

$$\mu_i = g_i^*(T,P^0) + \mathrm{R}T \ln \frac{f_i^0}{P^0} y_i \gamma_i^{(I)} = g_i^*(T,P^0) + \mathrm{R}T \ln \frac{f_i^0}{P^0} + \mathrm{R}T \ln y_i \gamma_i^{(I)} \qquad [8.14]$$

So, by comparison with expression [8.13], we deduce for the fugacity of the gas:

$$f_i = f_i^0 y_i \gamma_i^{(I)} = \varphi_i^0 y_i \gamma_i^{(I)} P \qquad [8.15]$$

We define the fugacity coefficient of the gas i in the mixture by:

$$\varphi_i(T,P,n_1,\ldots,n_i,\ldots n_N) = \frac{f_i}{P} \qquad [8.16]$$

The gas will tend towards a perfect behavior if the pressure tends towards zero, i.e.:

$$\lim_{P \to 0} \varphi_i = \lim_{P \to 0} \frac{f_i}{P} = 1 \qquad [8.17]$$

By comparison with equation [8.15], the fugacity coefficient of the mixture can be written:

$$\varphi = \varphi_i^0 y_i \gamma_i^{(I)} \qquad [8.18]$$

Generally, we can write:

$$RT \ln \varphi_i = \lim_{P^0 \to 0} \int_{P^0}^{P} \left(\overline{V}_i - \frac{RT}{P} \right) \mathrm{d}P \qquad [8.19]$$

This relation allows us to calculate the fugacity coefficient of a gas from the state equation of a mixture and consequently its fugacity and chemical potential using expressions [8.15] and [8.13].

There is an extremely simple calculation of the fugacity of a gas i, when the gas mixture forms a perfect solution of imperfect gases, because we obtain $\gamma_i^{(I)} = 1$, and the fugacity, from relation [8.15], becomes:

$$f_i = f_i^0 y_i = \varphi_i^0 y_i P \qquad [8.20]$$

This relation includes what we call the *Lewis equation*. This relation is often a very good approximation, especially beyond critical conditions, to calculate the fugacity of a gas mixture from only the fugacity of a pure gas under the same pressure as the total pressure of the mixture.

8.2. Characterizing gas mixtures

There are three ways to characterize a gas mixture:

– using a state equation of the mixture allows us to calculate the fugacity coefficients using expression [8.19];

– with a compressibility coefficient of the mixture to use relation [7.49] for the mixture or even apply to the mixture, a law of corresponding states (see section 7.3.4) but this needs to have defined and calculated a pressure and a critical temperature of the mixture as a function of composition. We can then use generalized charts;

– using a representation method of a solution by applying relation [8.14]. This method links the representation method of certain condensed molecular solutions since it must determine a coefficient of activity.

8.2.1. *Method of the state equations of gas mixtures*

Many equations of state, some of which we saw in Chapter 7, have been adapted to mixtures. In the following section we are going to assess the main ones.

8.2.1.1. *Van der Waals state equation*

The generalized Van der Waals equation with relation [7.1] can be used for a gas mixture by defining coefficients a^{mix} and b^{mix} of the mixture using two relations, [8.21] and [8.22], which we call mixing laws.

$$b^{mix} = \sum_{i=1}^{N} x_i b_i \qquad [8.21]$$

$$a^{mix} = \sum_{i=1}^{N} \sum_{j=1}^{N} x_j a_{ij} \qquad [8.22]$$

Relation [8.21] only involves co-volumes b_i of different pure gases. Relation [8.22] involves mixed terms, a_{ij}, relative to two components which can be taken as the geometric average of coefficients a_i relative to two corresponding pure gases, according to:

$$a_{ij} = \sqrt{a_i a_j} \qquad [8.23]$$

The state equation obtained has the same pitfalls as the Van der Waals state equation for the pure gas, i.e. that it is not very precise especially when it greatly differs from the perfect behavior of gases and in particular near the critical conditions.

8.2.2. *The Beattie–Bridgeman state equation*

An equation was generalized from the Beattie–Bridgeman equation [7.28]. It is established by defining different coefficients from those of pure gases according to expressions [8.24] and [8.25].

$$A_0^{mix} = \left(\sum_{i=1}^{N} x_i \sqrt{A_{0(i)}} \right)^2 \quad B_0^{mix} = \sum_{i=1}^{N} x_i B_{0(i)} \quad a^{mix} = \sum_{i=1}^{N} x_i a_i \qquad [8.24]$$

$$b^{mix} = \sum_{i=1}^{N} x_i b_i \qquad c^{mix} = \sum_{i=1}^{N} x_i c_i \qquad [8.25]$$

Of course, multiplying the number of constants makes the quality of this state equation better than that of the Van der Waals equation.

8.2.2.1. *Development of the virial*

We can also use a development of the virial of the mixture similar to equation [7.20] used in the case of a pure gas.

8.2.2.1.1. Definitions of the virial coefficients for the mixture

The second and third coefficients are defined, respectively, by the second and third coefficients of pure gases and the mixed terms $B_{2(ij)}$ and $B_{3(ijk)}$ relative to two or three components, according to:

$$B_2^{mix} = \sum_{i=1}^{N} \sum_{j=1}^{N} x_i x_j B_{2(ij)} \qquad [8.26]$$

$$B_3^{mix} = \sum_{i=1}^{N} \sum_{j=1}^{N} \sum_{k=1} x_i x_j x_k B_{3(ijk)} \qquad [8.27]$$

In the next section, we will see how to interpret the second coefficient of the virial, which is often sufficient, by developing the same method we used in section 7.4.3 for pure gases.

8.2.2.1.2. Calculating the second coefficient of the virial of a mixture

Using the first relation [5.16], we can extend the method we used in section 6.4.1 to mixtures of gases, A, B, C, etc., by superposing several quasi-chemical equilibria. We ultimately obtain:

$$\ln I_I = \sum_{(AB)} \frac{N_{(A)} N_{(B)}}{\sigma_{(AB)} V} B_{(AB)}(T) \qquad [8.28]$$

In this relation, the sum extended to (AB) means that taking all the possible couples of molecules and $\sigma_{(AB)}$ equals 2 if the molecules A and B are identical, and 1 if they are different. For example, for a binary compound, relation [8. 28] will take the form:

$$\ln I_I = \frac{N_A^2}{2V} B_{AA}(T) + \frac{N_A N_B}{V} B_{AB}(T) + \frac{N_B^2}{2V} B_{BB}(T) \qquad [8.29]$$

The coefficients B_{AA} and B_{BB} are defined by relation [7.76] and the mixed coefficient B_{AB} is defined by:

$$B_{AB}(T) = -\int_0^{r_\infty} 4\pi r^2 \left[\exp\left(-\frac{\varepsilon_{AB}}{k_B T}\right) - 1\right] dr \qquad [8.30]$$

The canonical partition function of an ensemble is therefore:

$$\ln Z_C = \sum_{(A)} \ln Z_{C(A)} + \ln I_I \qquad [8.31]$$

The sum is extended to all components of the mixture.

If we choose one mole of a binary compound, then:

$$N = N_A + N_B = N_a \qquad [8.32]$$

The molar fraction x of one of the two gases becomes the composition variable and therefore:

$$x = \frac{N_A}{N_a} \text{ and } 1 - x = \frac{N_B}{N_a} \qquad [8.33]$$

The second coefficient of the virial is defined by an extension of the definition [7.95], i.e.:

$$B_2(T) = N_a \sum_{(AB)} \eta_{AB} x_A x_B B_{(AB)}(T) \qquad [8.34]$$

The sum is extended to all possible binary compounds, AA, AB, BB, etc., and this time the coefficient η_{AB} equals 2 if both components are different and 1 if they are identical. So, in the case of the binary compound AB, relation [8.34] is written:

$$B_2(T) = N_a \left[x^2 B_{AA} + 2x(1-x) B_{AB} + (1-x)^2 B_{BB} \right] \quad [8.35]$$

This quadratic form in the variation of the second coefficient of the virial with the molar fraction of one of the gases is followed by a number of mixtures of binary gas compounds. Figure 8.1 shows the results for the hydrogen/helium, hydrogen/argon and helium/argon couples.

Some authors have defined the mixed parameter B_{AB} as the average of the simple parameters B_{AA} and B_{BB}, for example:

$$2B_{AB} = B_{AA} + B_{BB} \quad [8.36]$$

This expression is logical when the force field of both types of molecules is identical or at least very similar, which explains how such a relation nicely represents the results obtained with the oxygen/nitrogen mixture.

Other state equations, which we have found for pure gases, have also been extended to gas mixtures. This is the case with the Soave–Redlich–Kwong equation, given by relations [7.10], [7.11] and [7.12], and that of the Peng-Robinson equation, given by relations [7.15], [7.16] and [7.17]. We will give the corresponding mixing laws since these laws are mainly used when determining the coefficients of activity of solutions near to the triple point (see section 8.3).

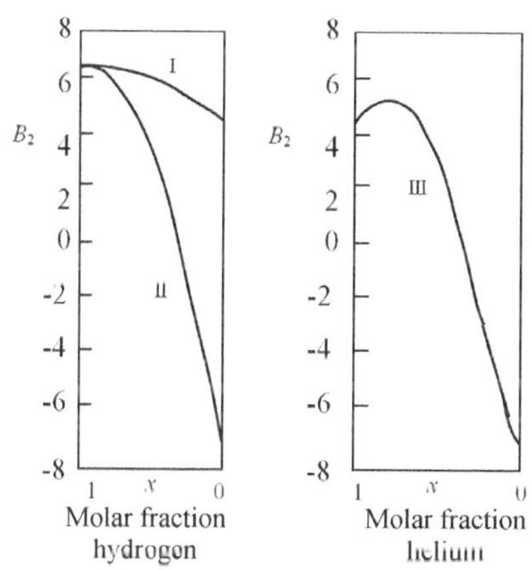

Figure 8.1. *Variations in the second coefficient of the virial with composition – I - H_2/He, II - H_2/Ar, III - He/Ar (from [GIB 29])*

8.2.3. *Calculating the compressibility coefficient of a mixture*

Calculating the compressibility coefficient of a mixture aims to use the same methods as those found when studying pure gases to quantify fugacities. There are two methods used to calculate the compressibility coefficients of a mixture. One uses the compressibility coefficients of pure gases and the other uses reduced diagrams which require the definition of relative variables (pressure and temperature) and therefore the corresponding critical variables.

8.2.3.1. *From coefficients of pure gases*

If we use relation [7.32] for each component of the mixture including partial pressure, i.e.:

$$V_m P_i = x_i Z_i \mathrm{R}T \tag{8.37}$$

The coefficient Z_i plays the same role as when the gas is pure since it measures how far the mixture is to perfection.

By summing all the terms relative to all the constituents, we obtain:

$$V_m \sum_{i=1}^{N} P_i = \mathrm{R}T \sum_{i=1}^{N} x_i Z_i \tag{8.38}$$

By keeping the same definition of partial pressure [8.4] as in the perfect solution of perfect gases, we can define a compressibility coefficient of the mixture by relation:

$$PV_m = Z^{mix} \mathrm{R}T \tag{8.39}$$

And we will obtain by comparing expressions [8.38] and [8.39]:

$$Z^{mix} = \sum_{i=1}^{N} x_i Z_i \tag{8.40}$$

This expression allows us to calculate the compressibility coefficient of the mixture from those of pure gases.

8.2.3.2. *Using reduced charts*

The second method used to obtain a compressibility coefficient of the mixture would be to apply the law of corresponding states and use generalized charts, provided that it has a reduced temperature and pressure, that is to say the temperatures and critical pressures of the mixture.

NOTE.– In a mixture, there is a triple point for each composition.

We are going to take a look at a binary mixture of two gases, A and B. In Figure 8.2, the two curves representing the liquefaction of two pure gases are shown up to the critical points A and B, as well as for two compositions, x_A and x'_A, the dew point and liquefaction curves. Each couple of these curves has a maximum at a point known as the critical point of the mixture. The geometric location of all these points is represented on the dotted line in Figure 8.2.

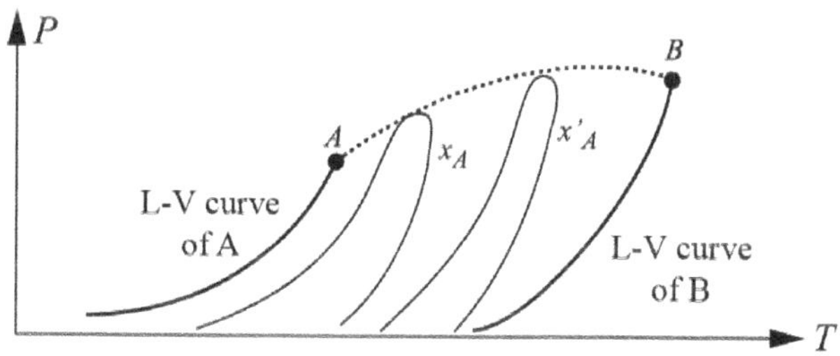

Figure 8.2. *Critical temperature and pressure of a binary compound*

Two methods are proposed to determine the critical point of a mixture.

The first method, proposed by Kay, assumes that all the critical points are aligned for the different compositions between the critical points of two pure compounds. The dotted line in Figure 8.2 is similar to a straight line. This method, for a mixture with N components, involves calculating the critical temperature and pressure of mixing by relations [8.41] and [8.42]:

$$P_c^{mix} = \sum_{i=1}^{N} x_i P_{c(i)} \qquad [8.41]$$

$$T_c^{mix} = \sum_{i=1}^{N} x_i T_{c(i)} \qquad [8.42]$$

However, when approaching the critical point of a mixture, as the line is not exactly straight, it is highly imprecise when determining whether we are ultimately dealing with one or two phases.

The second method, by Jaffré, proposes the empirical relations [8.43].

$$\begin{cases} \dfrac{T_c^{mix}}{\sqrt{P_c^{mix}}} = x_{\mathrm{A}}\phi_{\mathrm{A}}\sqrt{P_{c(\mathrm{A})}} + x_{\mathrm{B}}\phi_{\mathrm{B}}\sqrt{P_{c(\mathrm{B})}} \\ \dfrac{T_c^{mix}}{P_c^{mix}} = x_{\mathrm{A}}^2\phi_{\mathrm{A}} + x_{\mathrm{B}}^2\phi_{\mathrm{B}} + \dfrac{1}{4}x_{\mathrm{A}}x_{\mathrm{B}}\left[\phi_{\mathrm{A}}^{1/3} + \phi_{\mathrm{B}}^{1/3}\right]^3 \\ \phi_i = \dfrac{T_{c(i)}}{P_{c(i)}} \end{cases} \qquad [8.43]$$

These relations were tested on concrete cases, and allowed the critical pressure and temperature of a given mixture to be calculated, knowing those of the pure substances.

The reduced temperatures and pressures calculated from the critical data of pure gases allow us to determine, with the help of generalized charts used for pure gases, the compressibility coefficients of constituents of the mixture.

8.2.4. *Method using activity coefficients of solutions*

The last method developed to represent a mixture involves the use of relation [8.14] by superposing an expression of the activity coefficient deduced from a molecular model obtained using a model starting with the calculation of excess Gibbs energy as can be found for liquid solutions.

8.3. Determining activity coefficients of a solution from an equation of state

The methods, described above, used to determine a coefficient of activity from a model and the calculation of excess Gibbs energy, form part of a group of methods called *excess Gibbs energy methods* or G^{xs} *methods*. These methods give models that often find it difficult to represent vapor–liquid equilibria since these equilibria require, together with a description of the liquid phase, a reliable description of the gas phase.

Some authors inspired by the fact that "cubic" state equations, such as the van der Waals equation, account for the change in state from vapor–liquid and the critical point, and they have proposed to describe liquid and gas phases using the same model. These methods are called *state equation methods*.

8.3.1. *Methodology*

We will start with an expression giving the chemical potential of a component of a solution, with which we aim to determine the activity coefficient of this component in the solution.

For a real phase, which is an imperfect solution of non-perfect gases, we can write the chemical potential of one of the components in the form [8.13].

For the same imperfect gas, but assumed to belong to a perfect mixture, we express its chemical potential $\mu_i^{(\text{perf mix})}$ by expression [8.14].

The excess partial molar Gibbs energy of component i is the difference between the chemical potential in the real solution μ_i and its chemical potential in the perfect solution $\mu_i^{(\text{perf mix})}$ which we can write:

$$\overline{G_i^{xs}} = \mu_i - \mu_i^{\text{perf mix}} = \text{R}T \ln \gamma_i^{(I)} \qquad [8.44]$$

Therefore from the difference between expressions [8.13] and [8.14], we obtain:

$$\ln \gamma_i^{(I)} = \ln f_i - \ln f_i^0 x_i \qquad [8.45]$$

i.e.:

$$\gamma_i^{(I)} = \frac{f_i}{f_i^0 x_i} = \frac{\varphi_i}{\varphi_i^0 x_i} \qquad [8.46]$$

The fugacity coefficient of constituent i is given from the state equation using relation [7.46] and that of constituent i in the mixture can be calculated using expression [8.19], i.e.

$$\mathrm{R}T \ln x_i \gamma_i^{(I)} = \lim_{P^0 \to 0} \int_{\mathrm{P}_0}^{P} \left(\overline{V_i} - \frac{\mathrm{R}T}{P} \right) - \lim_{P^0 \to 0} \int_{\mathrm{P}_0}^{P} \left(V_i - \frac{\mathrm{R}T}{P} \right) \qquad [8.47]$$

Therefore, knowing a state equation of gases and the mixing laws of the parameters of this state equation helps obtain the activity coefficient of a component in the solution; this is so in the gas or liquid phase if this state equation can be applied to equally represent both the liquid phase and the gas phase.

This method was first applied by Van Laar who provided an expression of the activity coefficients using the Van der Waals state equation. Unfortunately, the success of the Van Laar model was limited by the fact that the Van der Waals equation, which applies to weakly imperfect gases, is very far from reality when approaching the liquid phase and even more so for the latter. The method was subsequently used with greater success with more efficient state equations accompanied by sophisticated mixing laws. In particular, this is the case for the PSRK model which we will describe.

NOTE.– The Van Laar law is, however, suitable for certain liquid alloys and also another model using an excess Gibbs energy method has been given.

8.3.2. *Studying solutions using the PSRK method*

The PSRK (Predictive Soave–Redlich–Kwong) method (1991) is based on the state equation of Soave, Redlich and Kwong (see section 1.3.2.3) with the mixing laws of parameters including the excess partial molar Gibbs energy calculated using a study method of liquid solutions called the *UNIFAC* method. The UNIFAC method considers that a solution of molecules, i, j etc., behaves like a solution with the functional groups, *k*, *m*, etc., which form the molecules. The activity coefficient of component *i* obeys the relation:

$$\ln \gamma_i^{(I)} = \ln \gamma_i^{(I)(\mathrm{Com})} + \ln \gamma_i^{(I)(\mathrm{Res})}$$

Its logarithm will be the sum of both contributions, one of conformation, with an entropic origin due to the arrangement molecules in the volume and the other residual, of enthalpic origin and due to the paired interactions between different molecules.

The UNIFAC model calculates these two contributions.

8.3.2.1. *The modified state equation of Soave–Redlich–Kwong and the mixing laws*

We will therefore use the Soave–Redlich–Kwong state function (relations [7.10] and [7.11]). The function α with an origin given by relation [7.12] has been modified and replaced, in the case of temperatures lower than the critical temperature by the Mathias and Coperman function, which takes the following form:

$$\alpha = \left[1 + C_1\left(1+\sqrt{T_r}\right) + C_2\left(1+\sqrt{T_r}\right)^2 + C_3\left(1+\sqrt{T_r}\right)^3\right]^2 \qquad [8.48]$$

In this expression, T_r is the reduced temperature, i.e. the ratio of the working temperature to the critical temperature, which obviously requires that the latter is known. C_1, C_2 and C_3 are three coefficients determined from the liquid–vapor equilibria of the pure substance. This function provides a better representation than function [7.12], which of course is kept in the absence of experimental data and if the temperature is greater than the critical temperature with the single term C_1.

The laws of mixing of the model are:

– for coefficient *b* the simple weighted additive rule, which means that we assume the volume of the mixture to be zero, we therefore write:

$$b^{mix} = \sum_i x_{\grave{a}} b_i \qquad [8.49]$$

– for coefficient a, the mixing law will use the excess Gibbs energy function, which we will determine for the liquid solution using the UNIFAC method:

$$\frac{a^{mix}}{RT}=\sum_i\left(x_i\frac{a_i}{b_iRT}-\frac{G^{xs}+\sum_i x_i\ln\frac{b_i}{x_i}}{0.64663}\right) \quad [8.50]$$

G^{xs} is the excess Gibbs energy which we will determine for the solution from UNIFAC data, that is to say the sum of the conformation and residual terms.

8.3.2.2. *Parameters of the model and overview of results*

The parameters of the model are therefore: the constants associated with the state equations of each component and those in expressions [7.10] and [7.11] including the parameters of equation [8.48], the critical temperatures and of each component and the data needed to use the UNIFAC model, i.e. following a good definition of the elementary groups chosen, their structural parameters and the energy parameters relative to each group couple.

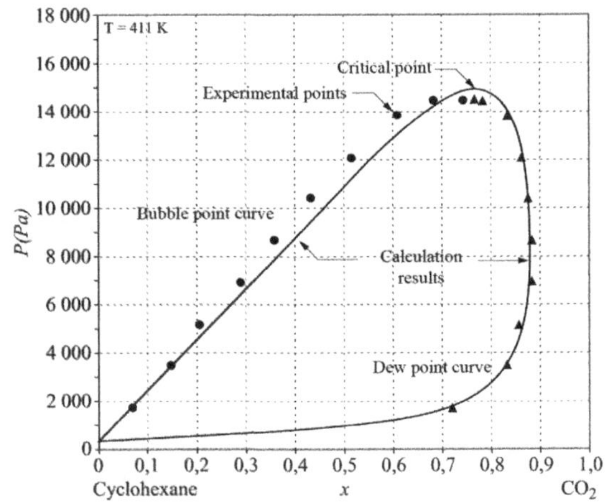

Figure 8.3. *Liquid-vapor chart of binary cyclohexane – carbon dioxide, comparison of experimental points and calculated curves*

The UNIFAC consortium (Dortmund Data Bank, DDB), together with the UNIFAC data bank, a specific PSRK data bank, provides the values of the ensemble of parameters for an increasing number of groups and substances; the data are collected from experimental studies on liquid-vapor equilibria.

By accurately defining the group chosen when using UNIFAC, this method gives accurate results, especially with regard to calculating liquid–vapor equilibria. The superiority of this method over the model is revealed when one component is in a supercritical condition and the critical point of the solution is reached. Figure 8.3 shows the goodness of fit of the experimental data and the values calculated in the case of the binary compound formed by cyclohexane and carbon dioxide both for the bubble point curve and for the dew point curve at 411 K.

8.3.3. *VTPR Model*

A certain number of downfalls for the PSRK model were noted by Ahler:

– determining the term α by relation [7.12] for temperatures that greatly exceed the critical temperature would quickly become insufficient;

– the volumic mass of the liquid phase would be poorly calculated by the SRK state equation [7.10];

– large errors would occur when there is a large variation in dimensions between the molecules of the components;

– the model would provide a poor prediction of the mixing enthalpy and activity coefficients of infinitely dilute components.

To address these pitfalls, the VTPR (Volume-Translated Peng–Robinson) model was proposed by Ahler and Gmehling in 2001.

It is based on the state function of Peng and Robinson (see section 7.3.2.4) modified by the translated volume and it takes the form:

$$P = \frac{RT}{V-t-b} - \frac{a\alpha}{(V+t)(V+t+b^2)+b(V-t-b)} \qquad [8.51]$$

t is the given translated volume, according to the critical conditions and compressibility coefficient under these critical conditions Z_c, by:

$$t = 0.242\frac{RT_C}{P_C}(Z_C - \varsigma) \text{ and } \varsigma = 2.5448Z_c - 0,4024 \qquad [8.52]$$

The function α is given by a Twu equation:

$$\alpha = \alpha_0(T_r) + \omega\left[\alpha_1(T_r) - \alpha_0(T_r)\right] \qquad [8.53]$$

ω is the coefficient of acentricity. $\alpha_0(T_r)$ is defined by:

$$\alpha_0(T_t) = T_r^{-0.201158}\exp\left[0.141599\left(1 - T_r^{2.29528}\right)\right] \qquad [8.54]$$

$\alpha_1(T_r)$ is defined by:

$$\alpha_1(T_t) = T_r^{-0.6601458}\exp\left[0.500315\left(1 - T_r^{2.633165}\right)\right] \qquad [8.55]$$

The rules of the mixture are:

– for t:

$$t^{mix} = \sum_i x_i t_i \qquad [8.56]$$

– for b:

$$b^{mix} = \sum_i \sum_j x_i x_j b_{ij} \text{ and } b_{ij} = \left(\frac{b_i^{3/4} + b_j^{3/4}}{2}\right)^{4/3} \qquad [8.57]$$

– for a:

$$a^{mix} = b\left(\sum_i x_i \frac{a_i}{b_i} - \frac{G_{\mathbf{Res}}^{xs}}{0.83087}\right) \qquad [8.58]$$

The excess Gibbs energy G^{xs}_{Res} is the residual term obtained from the *UNIFAC* (*DO*) model, used for liquid solutions and modified.

NOTE.– The difference in the excess Gibbs energy term selected with the *PSRG* method which was the complete variation in Gibbs energy and not only the residual term should be noted.

We find that the results obtained are similar to those of the *PSRG* model with regard to liquid-vapor equilibria (see Figure 8.4) but much better with regard to the weaknesses of this previous method and in particular for excess Gibbs energy (see Figure 8.5(a)) and the density of the liquid state and therefore the solid-liquid equilibria (see Figure 8.5(b)).

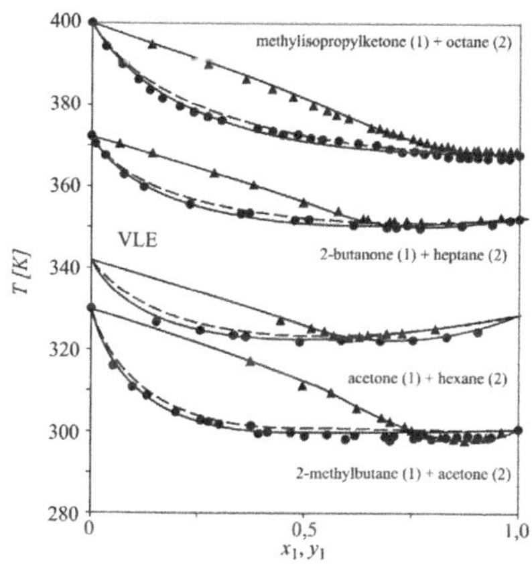

Figure 8.4. *Comparison between experimental data, PSRK results (---) and VTPR results (—) for some liquid-vapor equilibria (from Ahler and Gmehling, 2001)*

Unfortunately, the current VTPR method severely lacks available data sets that are specific and are still being accumulated by the Data Bank. In fact, the variations in the excess Gibbs energy term with temperature introduced in the mixing term [8.58] and those from the modified UNIFAC (Dortmund) model are too imprecise; the experimental curves from the data should be readjusted as Figure 8.6 shows, comparing this poor result from the VTPR model when using modified UNIFAC data and the good results given by VTPR with specially adapted data.

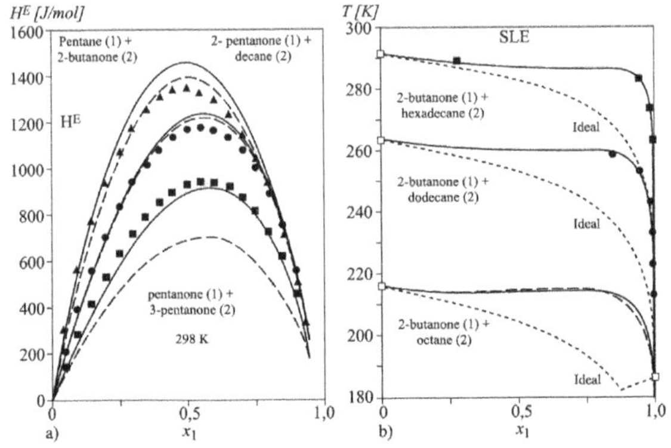

Figure 8.5. *Comparisons of experimental points, PSRK data and VTPR data for a) Excess Gibbs energy; b) solid-liquid equilibria (from Ahler and Gmehling, 2001)*

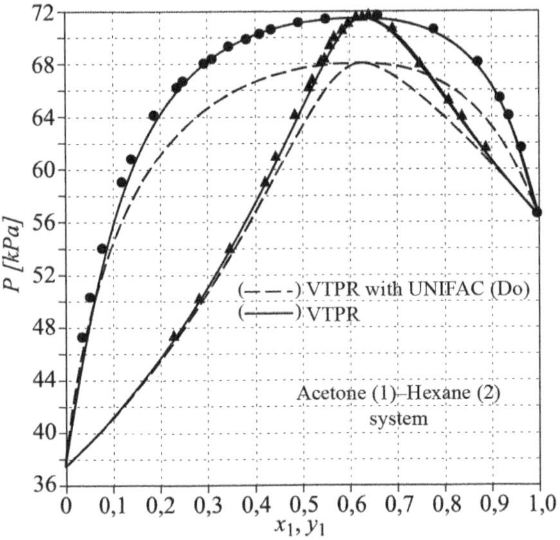

Figure 8.6. *Comparison of a liquid-vapor chart obtained experimentally, by VTPR and VTPR using UNIFAC (DO)data (according to Foster, Rarey, Ramjugernath, 2009)*

8.3.4. *VGTPR Model*

With regards to the issue indicated above with the VTPR method, researchers have tried to develop this method by using an enormous database available for modified UNIFAC Dortmund. It was found that the coefficients of activity from the VTPR model using the Dortmund data bank differed from those ultimately obtained by VTPR using G^{xs} UNIFAC *(Do)*:

$$G^{xs}_{VTPR} \neq G^{xs}_{G^{xs}UNIFACDo} \qquad [8.59]$$

which means a difference in each activity coefficient as shown in the example in Figure 8.7 obtained for the propane-benzene system, therefore:

$$\gamma_{i(VTPR)} \neq \gamma_{i(G^{xs}UNIFACDo)} \qquad [8.60]$$

This obviously results in significant differences, for example, in the liquid–vapor diagrams, and it seems that the activity coefficients obtained from the G^{xs} model are correct as shown in Figure 8.8 by the comparison between the curves obtained using the VTPR method using UNIFAC *(Do)* and those directly obtained with UNIFAC *(Do)* only for the liquid-vapor diagram of the propane-methanol binary compound.

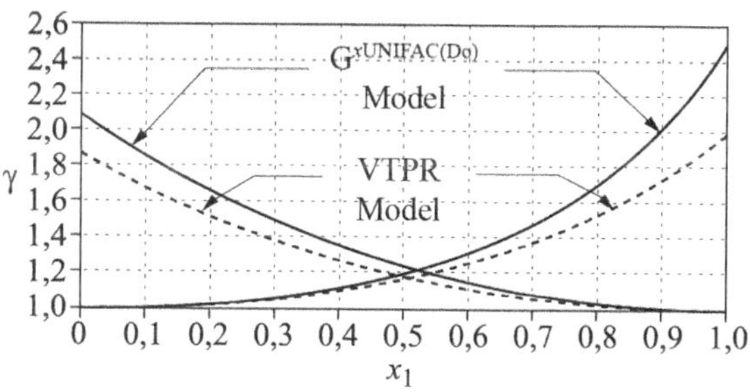

Figure 8.7. *Comparison of coefficients of activity calculated using both models in the case of binary propane-benzene (according to Foster, Rarey, Ramjugernath, 2009)*

The acronym VGTPR (Volume–Gibbs-Translated Peng–Robinson) (according to Ahler and Gmehling, 2001) suggests a modification which validates the G^{xs} model and will replace the value of $\gamma_{i(VOS)}$, obtained using the UNIFAC *(Do)* data bank, by the value $\gamma_{i(G^{xs}UNIFAC(Do))}$.

Here, an activity coefficient of transfer is introduced into the model, obtained using an iterative method.

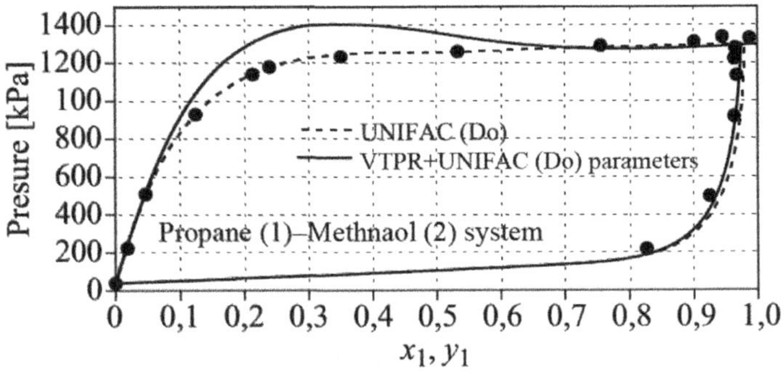

Figure 8.8. *Comparison between experimental data and VTPR data using UNIFAC (Do) and with UNIFAC (Do) only (according to Foster, Rarey, Ramjugernath, 2009)*

For this, the relation of mixture [8.58] is modified in the form:

$$a^{mix} = b\left(\sum_i x_i \frac{a_i}{b_i} - \frac{G_{\text{Res}}^{xs} + G_{\text{trans}}^{xs}}{0.83087}\right) \quad [8.61]$$

The excess Gibbs energy of transfer is calculated from:

$$G_{\text{trans}}^{xs} = \text{R}T \sum x_i \ln \gamma_{i(\text{trans})} \quad [8.62]$$

Each value $\gamma_{i(\text{trans})}$ is obtained using an iterative method according to the chart in Figure 8.9.

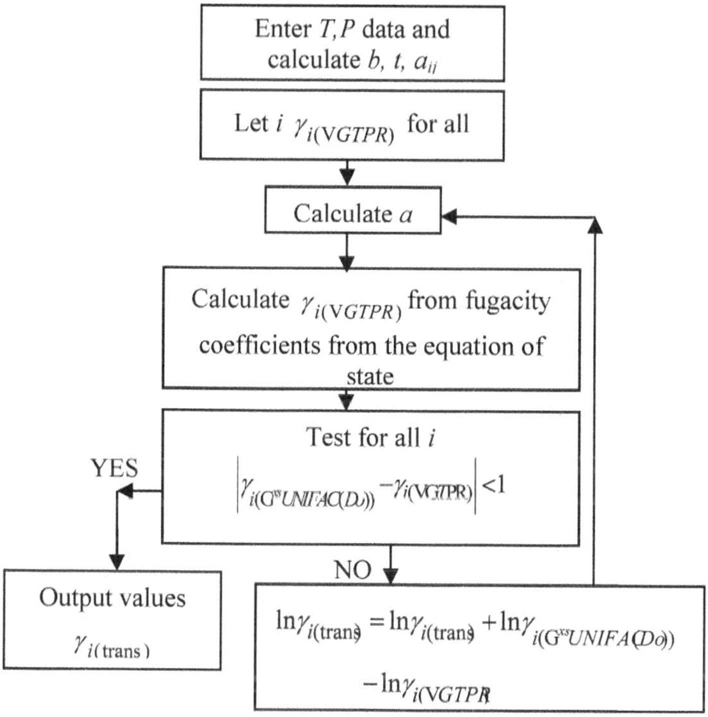

Figure 8.9. *Chart showing the iterations when calculating activity coefficients of transfer*

The results obtained are effectively quite encouraging and the amount of possible systems is infinitely greater than with the VTPR model since we have the whole DDB database available.

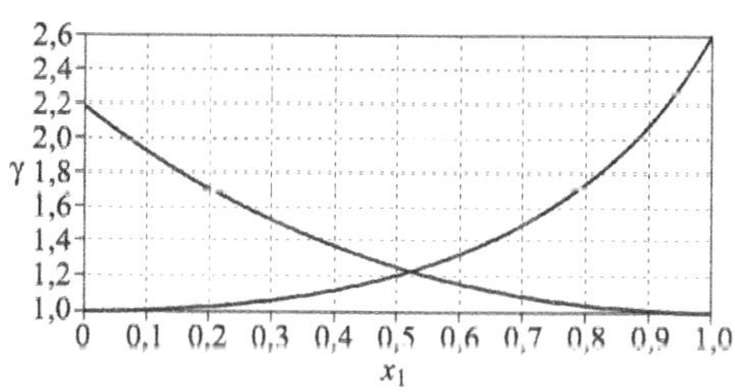

Figure 8.10. *Comparison of activity coefficients calculated using both VGTPR and UNIFAC models in the case of binary propane-benzene (according to Foster, Rarey, Ramjugernath, 2009)*

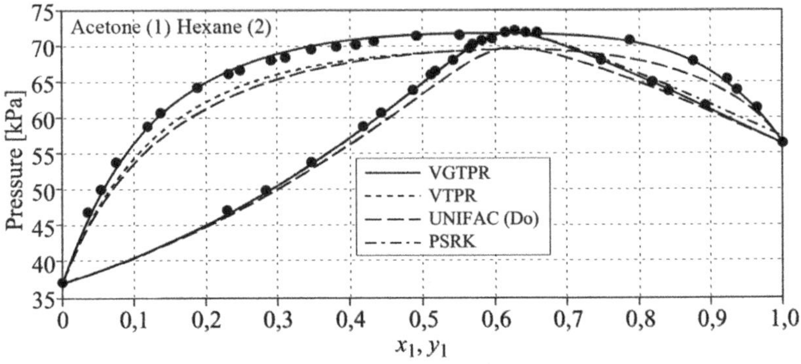

Figure 8.11. *Comparison of expérimental data with UNIFAC (Do) model, VTPR model, PSRK model and VGTPR model (according to Foster, Rarey, Ramjugernath, 2009)*

Appendices

Appendix 1

The Method of Lagrange Multipliers

A1.1. An overview of the problem

We aim to find the value of an extreme (maximum or minimum) of a function f of several variables x_1, x_2, x_n.

If the variables undergo small variations from $x_1 + \delta x_1$ to $x_i + \delta x_i$, the function varies from f to $f + \delta f$ such that:

$$\delta f = \sum_{i=1}^{n} \left(\frac{\partial f}{\partial x_i} \right) \delta x_i$$

For an extreme, $\delta f = 0$ and therefore:

$$\sum_{i=1}^{n} \left(\frac{\partial f}{\partial x_i} \right) \delta x_i = 0 \qquad \text{[A1.1]}$$

If all the variables are independent, this equation can be solved by cancelling out each term of the sum:

$$\left(\frac{\partial f}{\partial x_i} \right) = 0 \qquad \forall x_i$$

The problem that interests us here is to find the extreme when all variables x_i are not independent but where the relations are independent of one another, called constraints; the previous solution is therefore no longer valid. We will use the method of Lagrange multipliers.

A1.2. Solving the problem using the method of Lagrange multipliers

Let us assume, that to describe the method, there is a constraint between the variables x_i such that:

$$g(x_i) = 0$$

The constraint g is always true and the variable of g remains unchanged when the values of variables x_i vary, i.e:

$$\delta g(x_i) = \sum_{i=1}^{n}\left(\frac{\partial g}{\partial x_i}\right)\delta x_i = 0$$

Since δg is zero, we can multiply it by an arbitrary parameter λ and add it to equation [A1.1], we obtain:

$$\sum_{i=1}^{n}\left\{\left(\frac{\partial f}{\partial x_i}\right)+\lambda\left(\frac{\partial g}{\partial x_i}\right)\right\}\delta x_i = 0 \qquad [A1.2]$$

Equation [A1.2] can be solved using one of the variables, $i = n$ for example, as a function of the other variables (i = 1, 2, etc. n-1) which are independent. For the time being, λ is arbitrary, but we can choose its value so that the term in x_n in equation [A1.2] is zero, that is to say that we can select λ so that:

$$\left(\frac{\partial f}{\partial x_n}\right)+\lambda\left(\frac{\partial g}{\partial x_n}\right) = 0 \qquad [A1.3]$$

So equation [A1.2] becomes:

$$\sum_{i=1}^{n-1}\left\{\left(\frac{\partial f}{\partial x_i}\right)+\lambda\left(\frac{\partial g}{\partial x_i}\right)\right\}\delta x_i = 0$$

Now, the n-1 variables are independent and the solution is:

$$\left(\frac{\partial f}{\partial x_i}\right)+\lambda\left(\frac{\partial g}{\partial x_i}\right)=0 \qquad i=1,2,\ldots,n-$$

However, equation [A1.3] has exactly the same form and therefore the maximum or minimum of f can be found by solving the system:

$$\left(\frac{\partial f}{\partial x_i}\right)+\lambda\left(\frac{\partial g}{\partial x_i}\right)=0 \qquad i=1,2,\ldots,n$$

If there are several constraints, we will introduce as many multipliers as constraints into [A1.2] and deduce as many equations in the form of [A1.3] as constraints.

A1.3. Determining the multiplier values

There are two methods to determine the value of the arbitrary constant λ.

The first method involves solving equation [A1.3] rather than carrying out the optimization.

The second method involves keeping the undetermined multiplier λ until a property is deduced for which the value is known. This is the method that is often used in statistical thermodynamics to determine the multiplier $\beta = 1/\mathrm{k_B}T$.

In the case of several multipliers, the methods can be mixed, with some constants determined by the first method and others by the second. This is found when determining both constants α and β in statistical thermodynamics (see sections 4.2.2, 4.2.3, 4.5.3 and 4.5.4).

Appendix 2

Moments of Inertia of Molecules

In the tables below, we find the expressions of moments of inertia for several common forms of molecules.

m is the total mass of molecule, m_A is the mass of atom A.

Rotor	Form	I
Diatomic molecules	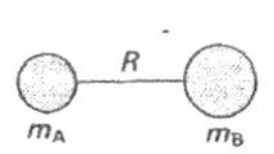	$\frac{m_A m_B}{m_A + m_B} R^2$
Linear rotor	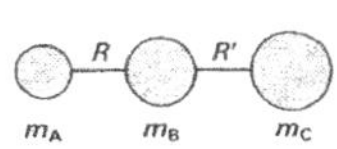	$m_A R^2 + m_C R'^2 - \frac{(m_A R - m_C R')^2}{m}$
Linear rotor	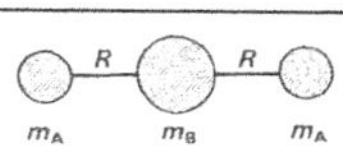	$2 m_A R^2$
Symmetrical rotor	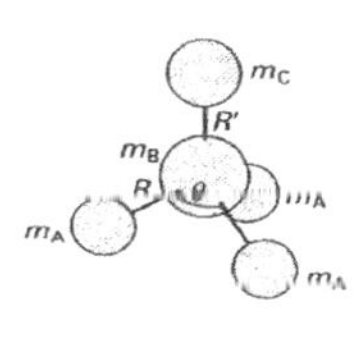	$I_{II} = 2 m_A R^2 (1 - \cos\theta)$ $I_{=} = m_A R^2 (1 - \cos\theta) + \frac{m_A}{m}(m_A + m_C) R^2 (1 + 2\cos\theta) + \frac{m_C R'^2}{m} \left\{ \begin{matrix} (3m_A + m_B) R' + \\ 6 m_A R \left[\frac{1}{3}(1 + 2\cos\theta) \right]^{1/2} \end{matrix} \right\}$

Rotor	Form	I
Symmetrical rotor	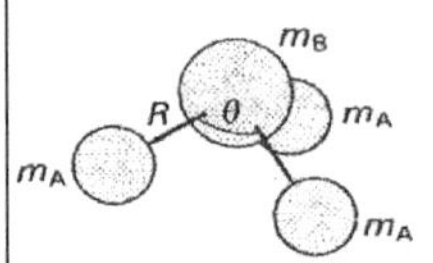	$I_{II} = 2m_A R^2(1-\cos\theta)$ $I_{\perp} = m_A R^2(1-\cos\theta)$ $+\dfrac{m_A m_B}{m} R^2(1+2\cos\theta)$
Symmetrical rotor	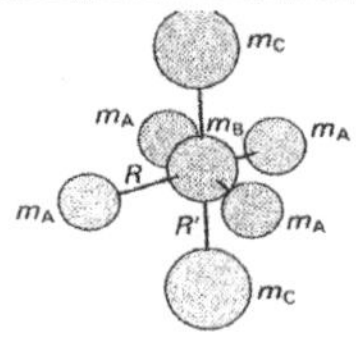	$I_{II} = 4m_A R^2$ $I_{\perp} = 2m_A R^2 + 2m_C R'^2$
Spherical rotor	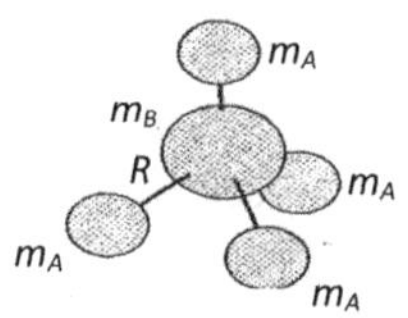	$\dfrac{8}{3} m_A R^2$
Spherical rotor	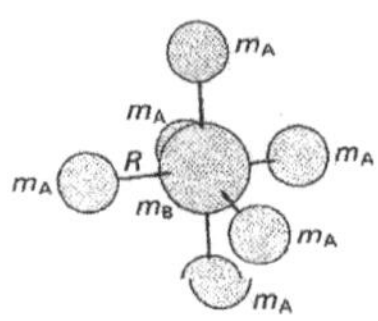	$4m_A R^2$

Appendix 3

Mathematical Complements

This section covers a number of mathematical rules used in this book.

A3.1. Homogeneous functions and Euler functions

We call a homogeneous function of degree m with regard to variables x, y and z the functions $f(x,y,z)$ so that for any independent variable k of x, y and z, we obtain:

$$f(kx,ky,kz)=k^m f(x,y,z) \qquad [A3.1]$$

EXAMPLES.– Homogeneous functions with a degree of zero are such that:

$$f(kx,ky,kz)=f(x,y,z) \qquad [A3.2]$$

Functions with a degree of 1 are such that:

$$f(kx,ky,kz)=kf(x,y,z) \qquad [A3.3]$$

It is easy to show that if a function $f(x,y,z)$ is homogeneous with a degree of m then the ratio $f(x,y,z)/x^m$ has a degree of zero.

$$\frac{f(kx,ky,kz)}{kx^m}=\frac{f(x,y,z)}{x^m} \qquad [A3.4]$$

Homogeneous functions obey Euler's identity:

$$\frac{x\partial f(kx,ky,kz)}{\partial(kx)}+\frac{y\partial f(kx,ky,kz)}{\partial(ky)}+\frac{z\partial f(kx,ky,kz)}{\partial(kz)}=mk^{m-1}f(x,y,z) \quad [A3.5]$$

This expression is clear after deriving relation [A3.1].

The reciprocal is also true: if a function *f(x,y,z)* obeys Euler's identity, it is homogeneous with a degree of *m* with regard to variables *x, y* and *z*.

EXAMPLES.– Homogeneous functions with a degree of zero will satisfy:

$$\frac{x\partial f(kx,ky,kz)}{\partial(kx)}+\frac{y\partial f(kx,ky,kz)}{\partial(ky)}+\frac{z\partial f(kx,ky,kz)}{\partial(kz)}=0 \quad [A3.6]$$

Functions with a degree of 1 satisfy:

$$\frac{x\partial f(kx,ky,kz)}{\partial(kx)}+\frac{y\partial f(kx,ky,kz)}{\partial(ky)}+\frac{z\partial f(kx,ky,kz)}{\partial(kz)}=f(x,y,z) \quad [A3.7]$$

These notions are useful in thermodynamics since many functions such as the characteristic functions, volume etc. are homogeneous with a degree of 1 with regard to the amount of matter, that is to say, the number of moles $n_1, n_2, \ldots$ The corresponding partial molar variables derived from the previous with regard to the amounts of matter are therefore homogeneous with a degree of zero with regard to these amounts of matter.

A3.2. Exact total differentials

A3.2.1. *Definition and identity of Maxwell*

Let *P* and *Q* be two continuous functions derived from the variables *x* and *y*. The expression, called Pfaff's expression, is written:

$$P(x,y)\mathrm{d}x+Q(x,y)\mathrm{d}y \quad [A3.8]$$

It is an exact total differential if it is the total differential of a function *u* of the same variables *x* and *y,* i.e.:

$$\mathrm{d}u=P\mathrm{d}x+Q\mathrm{d}y \quad [A3.9]$$

We therefore obtain:

$$\frac{\partial u}{\partial x} = P \text{ and } \frac{\partial u}{\partial y} = Q$$

The main property of exact total differentials is that they obey Maxwell's identity, which is written:

$$\frac{\partial P}{\partial y} = \frac{\partial^2 u}{\partial x \partial y} = \frac{\partial^2 u}{\partial y \partial x} = \frac{\partial Q}{\partial x} \text{ i.e. } \frac{\partial P}{\partial y} = \frac{\partial Q}{\partial x} \qquad [A3.10]$$

The previous equality is needed and is sufficient so that [A3.8] is an exact total differential.

A3.2.2. *Integrating factor*

If an expression such as [A3.8] is not an exact total differential, it is written in the form:

$$\delta u = P \mathrm{d}x + Q \mathrm{d}y \qquad [A3.11]$$

It is possible to transform this differential form into an exact total differential, and then integrate it, by multiplying it by the factor L, called the integrating factor which, in general, is a function of the same variables as P and Q. Therefore, according to [A3.10], we obtain:

$$\frac{\partial (LP)}{\partial y} = \frac{\partial (LQ)}{\partial x} \qquad [A3.12]$$

For example, in thermodynamics, we know that the differential of the amount of heat δQ :

$$\delta Q = C_p \mathrm{d}T + h \mathrm{d}P \qquad [A3.13]$$

is not an exact total differential, but by multiplying it by the inverse of temperature, the function d(Q/T) is an exact total differential. The inverse of temperature is an integrating factor for the amount of heat and then leads to entropy.

A3.3. Cross partial derivatives

In this section, we have mentioned a number of identities concerning partial derivatives.

If Q is a $f_1(x,y)$ function, we can write:

$$\left(\frac{\partial Q}{\partial y}\right)_x \left(\frac{\partial y}{\partial x}\right)_Q \left(\frac{\partial x}{\partial Q}\right)_y = -1 \qquad [A3.14]$$

By multiplying [A3.10] by [A3.14] and by eliminating y, we obtain:

$$\left(\frac{\partial P}{\partial Q}\right)_x = -\left(\frac{\partial y}{\partial x}\right)_Q \qquad [A3.15]$$

Similarly, if P is a $f_2(x,y)$ function, we can write:

$$\left(\frac{\partial Q}{\partial P}\right)_y = -\left(\frac{\partial x}{\partial y}\right)_P \qquad [A3.16]$$

Since P and Q are functions of x and y, it is possible to eliminate x and write $Q=F(P,y)$, for which we obtain:

$$\left(\frac{\partial Q}{\partial P}\right)_y \left(\frac{\partial P}{\partial y}\right)_Q \left(\frac{\partial y}{\partial Q}\right)_P = -1 \qquad [A3.17]$$

By removing $\left(\frac{\partial Q}{\partial P}\right)_y$ from this expression and by including [A3.16], we obtain:

$$\left(\frac{\partial y}{\partial P}\right)_Q = \left(\frac{\partial x}{\partial Q}\right)_P \qquad [A3.18]$$

The equations [A3.10], [A3.15], [A3.16] and [A3.18] are called cross differential equations [A3.9].

A3.4. Elements of combinatorial analysis

In this section, we are going to mention several definitions and formulas of combinatorial analysis used in statistical thermodynamics.

A3.4.1. *Arrangements*

We consider an ensemble E of n objects.

A3.4.1.1. *Definition*

We call the arrangement of p objects any ordered sequence of p objects in n objects.

The number of arrangements of p objects in n is denoted by the symbol A_n^p.

We must have $1 \leq p \leq n$ and if $p > n$ then $A_n^p = 0$.

Two arrangements of p objects are therefore different if they differ from one another by the nature of objects that compose them or by their order in the sequence.

If the same object can be found several times in an arrangement, we therefore say that it is an arrangement with repetition; on the other hand, if the same object only appears once in an arrangement, this is called a simple arrangement or a strict arrangement.

A3.4.1.2. *Arrangements with repetition*

When the same object can be found several times in an arrangement, the number of arrangements that are repeated is therefore:

$$\mathbf{A}_n^p = n^p \text{ with } 1 < p < n \qquad [A3.19]$$

A3.4.1.3. *Simple arrangements or without repetition*

When an object can only appear once in an arrangement, the number of simple arrangements is given by:

$$\mathbf{A}_n^p = \frac{n!}{(n-p)!} \text{ with } 1 \le p \le n \qquad [A3.20]$$

A3.4.2. *Permutations*

We consider an ensemble E of n objects.

A3.4.2.1. *Permutations without repetition*

We call a permutation of n different objects any sequence of n objects or any arrangement n to n of these objects.

Using relation [A3.20] the number of permutations of n objects is therefore:

$$\mathrm{P}_n = \mathrm{A}_n^n = \frac{n!}{(n-n)!} = n! \qquad [A3.21]$$

A3.4.2.2. *Permutations with repetitions*

In the case where there are several repetitions k of the same object taken from n objects, the number of possible permutations of n objects must be related to the number of permutations of k identical objects.

The number of permutations of n objects is therefore:

$$\mathbf{P}_n = \frac{n!}{k!} \qquad [A3.22]$$

These permutations of k objects are all identical and must only be counted once.

A3.4.3. *Combinations*

We consider an ensemble E of n objects.

A3.4.3.1. *Definitions*

We call a combination of n objects p to p, or combination of p objects taken from n, any ensemble of p objects in n without replacing or repositioning the order in which the objects p are taken.

A3.4.3.2. *Simple combinations and combinations without repositioning*

If an object can only be counted once, this is a simple combination or a combination without repositioning. Note this combination:

$$C_n^p = \text{ or sometimes} \begin{pmatrix} n \\ p \end{pmatrix}$$

We show that the number of such combinations is:

$$C_n^p = \frac{n!}{p!(n-p)!} \text{ with } 1 \leq p \leq n \qquad [A3.23]$$

If $n < p$ then $C_n^p = 0$.

A3.4.3.3. *Combinations with repetition*

If an object can be counted several times, these are combinations with repetition.

We show that the number of combinations of n objects p to p with repositioning returns to the number of combinations of n+p-1 objects p to p. This number is therefore given by:

$$C_{n+p-1}^p = \frac{(n+p-1)!}{p!(n-1)!} \qquad [A3.24]$$

A3.4.3.4. *Properties of combinations*

The combinations have a symmetry property. We show that this is the same as giving the combination of p chosen objects as n-p not chosen objects, i.e.:

$$C_n^p = C_n^{n-p} \qquad [A3.25]$$

We show that the combinations obey Pascal's rule:

$$C_{n-1}^{p-1} + C_{n-1}^p = C_n^p \qquad [A3.26]$$

This is what we call the compound combination formula.

Appendix 4

Constants and Units

Here we will cover the main units used, the conversion factors and the values of the main physical constants.

A4.1. The international system of units (SI)

We have consistently preferred the international system of units SI based on the meter (m) for length, kilogram (kg) for mass, second (s) for time, ampere (A) for intensity of current, kelvin (K) for temperature, and mole (M) for quantity of matter (Table A.4.1).

The derivatives of these fundamental units are also used, with specific names including joule (J) for work and energy, watt (W) for power, pascal (Pa) for pressure and the coulomb (C) for amount of charge.

Multiples and sub-multiples of the previous units are also used, namely nano (multiple:10^{-9}), micro (multiple:10^{-6}), milli (multiple:10^{-3}), centi (multiple 10^{-2}), hecto (multiple: 100), kilo (multiple: 1000).

We also find other units not in the SI system. The main ones are:

– For pressures:

- atmosphere (atm) (1013.25 hPa),
- millimeter of mercury (mm Hg) (133.33 Pa),
- bar (b) (10^5 Pa).

– For work or energy:
  - calorie(cal) (4.184 J),
  - electronvolt (eV) ($1.59.10^{-19}$ J),
  - erg (10^{-7} J).
– For length:
  - angström (Ä) (10^{-10} m),
  - micron (1μm or 10^{-6} m).

Note also the old imperial measurements still found in the literature:

1 inch = 2.54 cm	1 square inch = 0.4516 cm^2
1 foot = 30.48 cm	1 square foot = 929.03 cm^2
1 yard = 0.9144 m	1 cubic inch = 16.3872 cm^3
1 mile = 1.6093 km	1 pound = 0.4536 kg
1 nautical mile = 1.852 km	

Name	Symbol used	Value
Ideal gas constant	R	8.315 $J.mole^{-1}.K^{-1}$
Planck's constant	h	$6.62.10^{-34}$ J.s
reduced Planck constant	h/2π	$1.05.10^{-34}$ J.s
Avogadro's number	N_a	$6.02.10^{23}$ $mole^{-1}$
Elementary charge	e	$1.602.10^{-19}$ C
Electron mass	m_e	$0.91.10^{-30}$ kg
Proton mass	m_p	$1.67.10^{-27}$ kg
Boltzmann constant	k_B=R/N_a	$1.38.10^{-23}$ $J.K^{-1}$
Zero Celsius	0 C	273.15 K
Rydberg constant	$\Re$	109673731 m^{-1}
"Radius" of hydrogen atom	α_0	$5.29.10^{-11}$ m
Electrical permittivity of a vacuum	ε_0	$(1/36\pi).10^{-9} = 8.85.10^{-11}$ S.I
Magnetic permeability of a vacuum	μ_0	$4\pi.10^{-7} = 1.26.10^{-6}$S.I
Speed of light in a vacuum	c	3.10^8 $m.s^{-1}$
Relation: $\varepsilon_0\mu_0 c^2=1$		

Table A4.1. *Main physical constants*

A4.2. Electrostatic formulae

Table A.4.2 shows the main electrostatic formulae consistent in the international system; these are the relations used throughout this book.

Nature	Formula
Forces between point charges	$\overline{F} = \frac{1}{4\pi\varepsilon} \cdot \frac{qq'}{r^2}$
Field created by a point charge	$\overline{E} = \frac{1}{4\pi\varepsilon}.$
Gradient field	$\overline{E} = -\overline{\text{grad } V}.$
Voltage of asetof charges	$V = \frac{1}{4\pi\varepsilon} \cdot \sum \frac{q}{r}$
Electrical induction	$D = \varepsilon_0 E + P.$
Perfect dielectric	$\overline{D} = \varepsilon_0 (1 + \chi')E = \varepsilon \overline{E}.$
Electrical polarization	$\overline{P} = \chi' \varepsilon_0 \overline{E}$
Gauss's theorem	$\text{div } \overline{D} = \rho$
Poisson equation	$\Delta V + \frac{\rho}{\varepsilon} = 0$
Energy field	$W = \iiint \frac{\varepsilon E^2}{2} \text{dv}$
Rotational field	$\overline{\text{rot } E} = 0$
Notional distribution equivalent to the dielectric polarization	$\rho' = -\text{div } P$ and $\sigma' = P_n$

Table A4.2. *Main electrostatic formulae*

Appendix 5

Function Γ

A5.1. Definition

The function Γ (or *gamma function*), is the function of variable x, positive, defined by the integral:

$$\Gamma(x) = \int_0^{\infty} t^{x-1} \exp(-t) \mathrm{d} t \qquad [A5.1]$$

A5.2. Recurrence relation and table of values

By integrating expression [A5.1], it immediately becomes:

$$\Gamma(x+1) = x\Gamma(x) \qquad [A5.2]$$

This relation allows us to calculate the function Γ for any value of x when we know the values of Γ in an interval with an amplitude of 1. For example, Table A5.1 gives these values in the interval [1-2].

We find, by applying the expression of definition [A5.2], that if n is an integer then:

$$\Gamma(n+1) = n! \qquad [A5.3]$$

The function Γ of an integer is always confused with the factorial of the integer immediately before.

x	$\Gamma(x)$	x	$\Gamma(x)$	x	$\Gamma(x)$	x	$\Gamma(x)$
1.00	1.00000	1.25	.90640	1.50	.88623	1.75	.91906
1.01	.99433	1.26	.90440	1.51	.88659	1.76	.92137
1.02	.98884	1.27	.90250	1.52	.88704	1.77	.92376
1.03	.98355	1.28	.90072	1.53	.88757	1.78	.92623
1.04	.97844	1.29	.89904	1.54	.88818	1.79	.92877
1.05	.97350	1.30	.89747	1.55	.88887	1.80	.93188
1.06	.96874	1.31	.89600	1.56	.88964	1.81	.93408
1.07	.96415	1.32	.89464	1.57	.89049	1.82	.93685
1.08	.95073	1.33	.89338	1.58	.89142	1.83	.93969
1.09	.95546	1.34	.89222	1.59	.89243	1.84	.94561
1.10	.95135	1.35	.89115	1.60	.89352	1.85	.94561
1.11	.94739	1.36	.89018	1.61	.89468	1.86	.94869
1.12	.94359	1.37	.88931	1.62	.89592	1.87	.95184
1.13	.93993	1.38	.88854	1.63	.89724	1.88	.95507
1.14	.93642	1.39	.88785	1.64	.89684	1.89	.95838
1.15	.93304	1.40	.88726	1.65	.90012	1.90	.96177
1.16	.92980	1.41	.88676	1.66	.90167	1.91	.96523
1.17	.92670	1.42	.88636	1.67	.90330	1.92	.96878
1.18	.92273	1.43	.88604	1.68	.90500	1.93	.97240
1.19	.92088	1.44	.88580	1.69	.90678	1.94	.97610
1.20	.91817	1.45	.88565	1.70	.90864	1.95	.97988
1.21	.91558	1.46	.88560	1.71	.91057	1.96	.98374
1.22	.91311	1.47	.88563	1.72	.91258	1.97	.98768
1.23	.91075	1.48	.88575	1.73	.91466	1.98	.99171
1.24	.90852	1.49	.88595	1.74	.91683	1.99	.99581
						2.00	1.0000

Table A5.1. *Values of function $\Gamma(x)$*

A5.3. Extension to negative values

Relation [A5.2] can be written:

$$\Gamma(x) = \frac{\Gamma(x+1)}{x} \quad [A5.4]$$

This expression only has any sense if variable x is positive since it is the condition of definition.

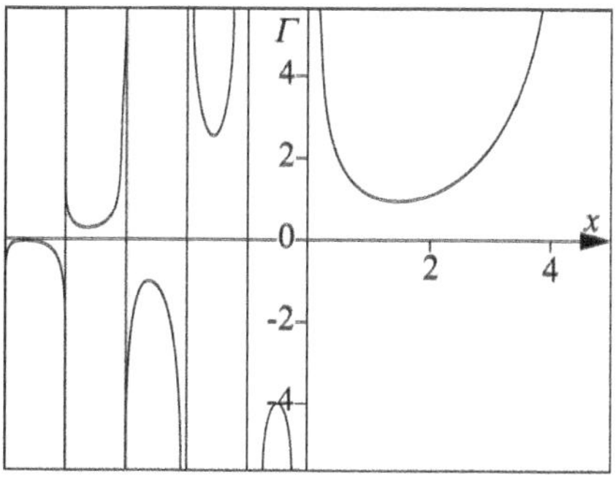

Figure A5.1. *Variations in function Γ(x)*

Now let us assume that variable x is between 0 and -1. $\Gamma(x+1)$ is therefore defined using relation [A5.4] and this relation [A5.4] can therefore be used to extend the definition of function $\Gamma(x)$ to the values of x between -1 and 0.

So, ultimately, we can define the function $\Gamma(x)$ by:

– the integral [A5.1] if x is positive;

– the formula of recurrence [A5.2] if x is negative.

So the function $\Gamma(x)$ can be defined by the formula of recurrence for all negative values with the exception of:

– $x = 0$;

– the negative integers $n < 0$.

If x tends towards zero, $\Gamma(x)$ tends towards infinity for positive values. If x tends towards -1, $\Gamma(x)$ tends towards infinity for negative values and so on.

Figure A5.1 represents the variations in function $\Gamma(x)$.

In particular, we show that:

$$\Gamma(-n-\frac{1}{2}) = \frac{(-1)^{2n+1} n!}{(2n+1)!}\sqrt{\pi} \qquad \text{[A5.5]}$$

$$\Gamma(n+\frac{1}{2}) = \frac{(2n)!}{2^{2n}\,n!}\sqrt{\pi} \qquad [A5.6]$$

x	1	1/2	-1/2
$\Gamma(x)$	0	$\sqrt{\pi}$	$-2\sqrt{\pi}$

Table A5.2. *Notable values of function $\Gamma(x)$*

Bibliography

AHLERS J.G., GMEHLING J., "Development of an universal group contribution equation of state I", Fluid Phase Equilibria, vol. 191, pp. 177-188, 2001.

ATKINS P.W., *Physical Chemistry*, Oxford University Press, Oxford 1990.

BEN-NAIM A., *Entropy and the Second Law*, World Scientific Publishing Company, New Jersey, 2012.

BERNACHE-ASSOLLANT D., COURNIL M., *Equilibres en solution*, Masson, Paris, 1997.

BESSON J., *Précis de thermodynamique et de cinétique électrochimiques*, Ellipses, Paris, 1984.

DEFAY R., PRIGOGINE I., *Tension superficielle et adsorption*, Edition Desoer, Liège, 1951.

DEMANGE H., GERMAIN G., NOTIN M., *Comprendre et appliquer la thermodynamique*, Masson et Cie, Paris, 1973.

DENBIGH K., *The Principles of Chemical Equilibrium*, Cambridge University Press, Cambridge, 1968.

DESRE P., HODAJ F., *Thermodynamique des matériaux – Equilibres de phase et métastabilité*, EDP Sciences, Paris, 2010.

EMSCHWILLER G., *Chimie Physique*, P.U.F., Paris, 1951.

FOSTER A., RAREY J., RAMJUGERNATH D., An improved group contribution volume translated Peng-Robinson equation of state – VGTPR, available at: www.ddbst.com/files/files/Durban/SACEC2009-Foster-VGTPR_Poster.pdf, 2009.

FOWLER R., GUGGENHEIM E.A., *Statistical Thermodynamics*, Cambridge University Press, London, 1949.

GERMAIN J.E., *Catalyse hétérogène*, Dunod, Paris, 1959.

GIBBY C.W., TANNER C.C., MASSON I., "The pressure of gaseous mixtures II. Helium and hydrogen and their intermolecular forces", *Proceedings of the Royal Society of London A*, vol. 122, pp. 283 – 304, 1929.

GILBERT J., *Chimie–Physique I - Atomistique et liaison chimique*, Masson, Paris 1963.

GLASSTONE S.N., LAIDLER K.J., EYRING H.W., *The Theory of Rate Processes*, MacGraw Hill Book Company, New York, 1941.

GOKCEN N.A., REDDY R.G., *Thermodynamics*, Plenum Press, New York and London, 1996.

GREINER W., *Mécanique quantique – une introduction*, Springer, Berlin, 1999.

HILL T.L., *Statistical Mechanics*, MacGraw-Hill, New York, 1956.

INFELTA P., GRAETZEL M., *Thermodynamique, Principes et Applications,* Brown Walker Press, 2006

KLOTZ I.M., ROSENBERG R.W., *Chemical Thermodynamics Basic Concepts and Methods*, 7th edition, John Wiley & Sons, 2008.

KRÖGER F.A., *The Chemistry of Imperfect Crystals*, North Holland Publishing Company, Amsterdam, 1974.

LALAUZE R., *Physico-chimie des interfaces solide-gaz 1*, Hermes-Lavoisier, Paris, 2006.

LE GORREC B., Thermodynamique et cinétique électrochimique expérimentale, available at: minatec.grenoble-inp.fr/Cours-A.pdf, 2005.

MARCHAND A., PACAULT A., *La thermodynamique mot à mot*, De Boeck University, Brussels, 1995.

MCQUARRIEN D.A., SIMON J.D., *Molecular Thermodynamics*, University Science Book, 1999.

METZGER E., "Loi de variation de la tension superficielle avec la temperature", *Journal de Physique et Le Radium*, vol. 7, no. 10, pp.303-309, 1946. Available at: http://dx.doi.org/10.1051/physrad.01946007010030300.

PACAULT A., *Eléments de thermodynamique statistique*, Masson, Paris, 1963.

PRIGOGINE I., DEFAY R., *Thermodynamique chimique*, Dunod, Paris, 1946.

PRINCE A., *Alloy Phase Equilibria*, Elsevier Publishing Company, Amsterdam, London, New York, 1966.

REID R.C., PRAUSNITZ J.M., POLING B.E., *Properties of Gas and Liquids*, MacGraw-Hill, New York, 1987.

RENON H., *Calcul sur ordinateur des équilibres liquide–vapeur et liquide–liquide*, Editions Technip, Paris, 1971.

SORENSEN O.T., *Non-stoichiometric Oxides*, Academic Press, New York, 1981.

SOUCHAY P., *Thermodynamique chimique*, Masson, Paris, 1968.

SOUSTELLE M., GARDET J.J., GUILHOT B., *Diagrammes de phases et stœchiométrie*, SUCHET J.P. (ed.), p. 33-50, Masson, Paris 1973.

SOUSTELLE M., *Modélisation macroscopique des transformations physico-chimiques*, Masson, Paris, 1990.

SOUSTELLE M., "Défauts ponctuels dans les solides in chimie", *Chimie–physique du frittage*, BERNACHE–ASSOLLANT D. (ed.) Hermes, Paris, 1993.

SOUSTELLE M., *Handbook of Heterogeneous Kinetics*, ISTE Ltd, London and John Wiley & Sons, New York, 2010.

SOUSTELLE M., *Cinétique chimique: éléments fondamentaux*, Hermes–Lavoisier, Paris, 2011.

Index

A, B, C

D, E

F, G, H

I, K, L

M, N, O

P

Other titles from

ISTE

in

Chemical Engineering

2014

DAL PONT Jean-Pierre, AZZARO-PANTEL Catherine
New Approaches to the Process Industries: The Manufacturing Plant of the Future

2012

DAL PONT Jean-Pierre
Process Engineering and Industrial Management

2011

SOUSTELLE Michel
An Introduction to Chemical Kinetics

2010

SOUSTELLE Michel
Handbook of Heterogenous Kinetics

Printed and bound by CPI Group (UK) Ltd, Croydon, CR0 4YY

13/09/2026

14961020-0003